Helical and Spiral Antennas —A Numerical Approach

An antenna system for direct reception of the broadcasting satellite TV-programmes. A backfire helical antenna is used as a primary feed for a parabolic reflector antenna. (courtesy of TDK CORPORATION)

The Author

Hisamatsu Nakano was born in Ibaraki, Japan, on April 13, 1945. He received the Ph.D. degree in electrical engineering from Hosei University, Tokyo, Japan, in 1974.

Since 1973 he has been on the faculty of Hosei University, where he is presently Professor of Electrical Engineering. For the period of March to September in 1981, he was Visiting Associate Professor at the University of Syracuse, New York, and contributed to formulating an electromagnetic coupling between a wire and a slot. For the period of February 1986 to September 1986, he was Visiting Professor at the University of Manitoba, Winnipeg, Canada. During his stay in Canada, he received International Scientific Exchange Award. From September 1986 to March 1987, he was Visiting Professor at the University of California, Los Angeles, and succeeded in analysing printed spiral antennas. His contribution entitled "Spiral slot antenna" was awarded a prize for the best paper in The Fifth IEE International Conference on Antennas and Propagation (ICAP 87).

ELECTRONIC & ELECTRICAL ENGINEERING RESEARCH STUDIES

ELECTROMAGNETIC APPLICATIONS SERIES

Series Editor: **Dr Ann Henderson,** *Royal Military College of Science, Shrivenham, Swindon, England*

1. Physical Techniques in Clinical Hyperthermia
   Edited by **Jeffrey W. Hand** *and* **James R. James**

2. Helical and Spiral Antennas—A Numerical Approach
   Hisamatsu Nakano

Helical and Spiral Antennas —A Numerical Approach

Professor Hisamatsu NAKANO

Department of Electrical Engineering
College of Engineering, Hosei University
Tokyo, Japan

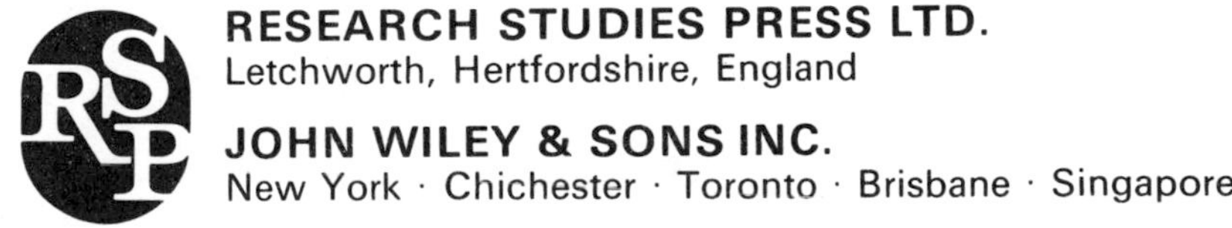

RESEARCH STUDIES PRESS LTD.
Letchworth, Hertfordshire, England

JOHN WILEY & SONS INC.
New York · Chichester · Toronto · Brisbane · Singapore

RESEARCH STUDIES PRESS LTD.
58B Station Road, Letchworth, Herts. SG6 3BE, England

Marketing and Distribution:

Australia, New Zealand, South-east Asia:
Jacaranda-Wiley Ltd., Jacaranda Press
JOHN WILEY & SONS INC.
GPO Box 859, Brisbane, Queensland 4001, Australia

Canada:
JOHN WILEY & SONS CANADA LIMITED
22 Worcester Road, Rexdale, Ontario, Canada

Europe, Africa:
JOHN WILEY & SONS LIMITED
Baffins Lane, Chichester, West Sussex, England

North and South America and the rest of the world:
JOHN WILEY & SONS INC.
605 Third Avenue, New York, NY 10158, USA

Library of Congress Cataloging-in-Publication Data

Nakano, Hisamatsu, 1945–
Helical and spiral antennas.

(Electronic & electrical engineering research studies. Electromagnetic applications series; 2)
Bibliography: p.
Includes index.
1. Antennas (Electronics). 2. Numerical analysis.
I. Title.
TK7871.6.N34 1987 621.38'0282 87-16667

ISBN 0 471 91736 2 (U.S.)

British Library Cataloguing in Publication Data

Nakano, Hisamatsu
Helical and spiral antennas: a numerical approach.—(Electronic & electrical engineering research studies. Electromagnetic applications series: no. 2).
1. Radio—Antennas
I. Title II. Series
621.3841'35 TK6565.A6

ISBN 0 86380 060 2

ISBN 0 86380 060 2 (Research Studies Press Ltd.)
ISBN 0 471 91736 2 (John Wiley & Sons Inc.)

Printed in Great Britain by Galliard (Printers) Ltd., Great Yarmouth

Editorial Preface

Helical and spiral antennas with their wide bandwidth characteristics have always been a fascinating topic, but in recent years the system demands on the performance of these antennas have become more stringent and complex. Professor Nakano of Hosei University in Japan has specialised in spiral-type radiators for several years and has now compiled a most comprehensive and detailed analytical account of their behaviour, which I am pleased to present to you as the second book in my Series on Electromagnetic Applications. The topic of retaining good circular polarisation over a very wide bandwidth is a particularly interesting feature of this book and is a technical problem which I'm sure will be familiar to many of you.

The research presented by Professor Nakano is outstanding for the success in which numerical techniques have been applied to this class of antenna which to date has defied precise analysis. The comprehensive approach presented lays an excellent foundation for the study of many other types of spiral elements such as cavity-backed spirals which are commonly used in practical systems, and is an important contribution to our understanding of these antennas.

ANN HENDERSON
Series Editor

All mankind are like grass, and all their glory is like wild flowers. The grass withers, and the flowers fall but •••••

To Mitsuko and Misae

Acknowledgments

The author would like to thank the Editor of the Research Studies on Antennas, Professor J. R. James, for his encouragement to write this monograph. To the Editor of Electromagnetic Applications Series, Dr. A. Henderson, the author is particularly grateful for her valuable advice and suggestions during the preparation of the manuscript. The author also acknowledges the help of Dr. J. Yamauchi who contributed to reviewing each chapter. The same refers to Professor L. Shafai, Mr. K. Hirose, Miss L. S. Solomon and Mr. E. Medina Jr.

Finally, thanks are due to the graduate students for typing the manuscript and drafting illustrations with great diligence:

Mr. T.Iwasaki(Sects.1.1-1.3), Mr. K.Yabe(Sects.1.4-1.5),
Mr. T.Ikekawa(Sects.1.6 and 9.3), Mr. S.Arai(Chapt.2),
Mr. M.Eda(Chapt.3), Mr. M.Tanabe(Chapt.4),
Mr. H.Mimaki(Chapt.5), Mr. Y.Kojima(Sects.6.1-6.4),
Mr. Y.Samada(Sects.6.5-6.8), Mr. K.Yamaki(Chapt.7),
Mr. M.Sato(Chapt.8), and Mr. Y.Minegishi(Sects.9.1-9.2).

Preface

This monograph is the summary of the past five years of research investigations made in The Antenna Laboratory at Hosei University. The subjects described in the monograph are antennas and scatterers for a circularly polarised wave. It is the purpose of the monograph to present the radiation characteristics of these antennas and scatterers with emphasis on their engineering aspects. The author presumes that the reader has a knowledge of basic electromagnetic theory, including Maxwell's equations, integral equations of wire antennas, and the calculation of radiation characteristics.

The monograph covers nine associated topic areas:

(1) numerical method and technique for antenna analysis
(2) square spiral antenna
(3) two-wire round spiral antenna
(4) spiral antenna with two off-centre sources
(5) polarisation diversity of spiral antenna
(6) helical antenna of endfire mode
(7) helical antenna of backfire mode
(8) conical helix antenna
(9) wire scatterers

which have for the first time been brought together in a fully descriptive account, and as such constitutes a unique contribution to the world literature.

Chapter 1 describes how to determine the current distributions on conducting wires. The moment method is briefly summarised, and an integral equation for an arbitrary configuration is viewed from the angle of the point-matching method. The author derives a simplified kernel, aiming at easier treatment of the integral equation. In the rest of the chapter, a pair of integral equations for a system consisting of wires and slots are presented, and some comments on the calculation for radiation characteristics are added.

Chapter 2 deals with small and large square spiral antennas. In the small spiral, the input impedance is obtained as a function of the wire radius. In addition, the axial ratio when the antenna plane is bent is evaluated on the basis of the theoretically determined radiation field. In the large spiral antenna, the mechanism of the wideband radiation characteristics is explained from the behaviour of the current distribution.

Chapter 3 is regarded as the counterpart of Chapter 2. It covers the radiation characteristics of a two-wire round spiral antenna with a single voltage source. The outermost arms of the spiral are transformed into small zigzag elements. The suppression of the reflected current along the zigzag elements leads to superiority in the axial ratio over a smooth spiral antenna without zigzag elements.

Chapter 4 is concerned with the polarisation diversity of a two-wire spiral antenna. The antenna is different from the previous antennas in Chapters 2 and 3 in that it has two voltage sources which are located at the two points symmetrical with respect to the spiral origin. The concept of the active region for the radiation is introduced, and the rotational sense of circular polarisation is investigated in detail.

This study of polarisation diversity is followed by Chapter 5, where a spiral antenna of four arms is surveyed for two cases; (a) the two arms are driven and the remaining arms are parasitic, (b) all the four arms are driven. The current distributions for the two cases are used for interpreting the polarisation diversity of the radiation field. The author refers to the effects of the parasitic arms on the radiation field.

Chapter 6 is devoted to analysing a helical antenna of

endfire mode. The model antenna analysed is a balanced-type helical antenna. It is noted that the model antenna may be used as a tool for detecting a circularly polarised wave of right-handed sense and one of left-handed sense of an unknown incident wave. The analysis of this chapter provides theoretical information on the phase velocity of the current on the helix, and clarifies the effects of tapering the open end section on the axial ratio. In the remainder of the chapter a short helical antenna is investigated, and an array consisting of short helices is proposed as a practical example of circular polarisation.

Chapter 7 begins with a comparison of the radiation characteristics of three types of backfire helical antennas. The front to back (F/B) ratios are examined with special interest. Subsequently, the author analyses a backfire helical antenna terminated with a resistive load, and demonstrates that the F/B ratio is improved. The location of the phase centre is also examined for the use of a primary feed antenna for a paraboloidal reflector.

In Chapter 8 a conical helix antenna is studied from the viewpoint of widening a frequency band. This chapter is recognised as an extension of Chapter 6. The investigation of the current distribution along a long antenna arm reveals that the minimum length of the arm for wide-band operation exists. In the rest of the chapter a helix antenna of low silhouette is analysed, and the bandwidth is determined by the radiation characteristics.

Chapter 9 deals with scatterers for a circularly polarised wave. The scatterers considered are helical and crossed wires. When a circularly polarised wave illuminates these two scatterers, the former acts as a directive element of the circularly polarised wave, while the latter acts as a reflective element, whose performance is the same as that of a dihedral corner reflector. It is, therefore, noted that the helical wires may be used for enhancing an antenna gain or controlling an antenna gain. An example of the power enhancement is demonstrated with a square spiral antenna. It is also noted that the crossed wire may be used as a back-scattering element of circular polarisation. This is verified by experimental work with arrays consisting of some crossed wires.

The author hopes that practising engineers and graduate students who are concerned with radio communication between ground stations and mobiles or aircraft, radar, electromagnetic compatibility, or related topics, will find this monograph to be inspiring research literature in these areas.

Hisamatsu NAKANO

Contents

Chapter 7. Helical Antenna of Backfire Mode

Chapter 8. Conical Helix Antenna

Chapter 9. Wire Scatterer for a Circularly Polarised Wave

CHAPTER 1
Numerical Method and Technique for Antenna Analysis*

1.1. INTRODUCTION

When designing wire antennas it is a matter of the greatest importance to obtain their current distributions, principally because the radiation characteristics can be evaluated directly from them. The current distribution of a wire antenna can be determined either experimentally or by numerical analysis. In the former case, there is a limit to the frequency at which accurate measurements can be made due to the field disturbances caused by the introduction of electrical probes into the near-field of the antenna. In the latter case, however, there are no limits to the frequency at which current distributions can be estimated and the accuracy is limited only by computational aspects or the model chosen. Our emphasis in this chapter, therefore, is put on some theoretical techniques which can give us reasonably accurate numerical results for current distributions.

There are two types of integral equations for determining the current distributions of wire antennas [1][2]. Each type of integral equations may be reduced to a set of simultaneous linear algebraic equations, or symbolically expressed in a matrix form [Z][I]=[V], where [Z], [I], and [V] are referred to as

* The work in this Chapter is taken from papers (p-2), (p-10), (p-14), (p-23), (c-12), (c-17), (c-18), and (c-19) listed at the end of this monograph. Other references, indicated in square brackets in the text, are listed at the end of the Chapter.

generalised impedance, current, and voltage matrices, respectively. Once the matrices [Z] and [V] are calculated, the matrix [I], i.e., the unknown current, is determined. Since the calculation of [Z] is more difficult than that of [V], we should pay special attention to the treatment of the generalised impedance matrix.

In Section 1.2 we study how to solve Pocklington's equation by using the moment method [3]. A piecewise sinusoidal Galerkin method is mentioned [4][5], and the generalised impedance matrices for a straight wire and an arbitrarily bent wire are presented in Section 1.3. It is noted that each matrix element of [Z] is given by a single integration.

In Section 1.4 we refer to Mei's integral equation for arbitrary wire antennas [6]. This equation, which is regarded as an extension of Hallen's integral equation for a straight wire [2], is solved by using a point-matching method (one of the moment methods). Subsequently, we derive Nakano's integral equation in Section 1.5, making use of Mei's integral equation. The derived kernel contains neither derivatives nor integrals, and hence it contributes to easier treatment of the integral equation for arbitrary wire antennas, with less computational time [7][8].

In Section 1.6 we study a pair of integral equations which are able to handle a system composed of some wires and slots [9]. Each kernel of the integral equations is expressed in simple algebraic form. It will be found that the results obtained in Section 1.5 constitute a part of the integral equations presented in this section.

1.2. INTEGRAL EQUATION AND THE MOMENT METHOD

It is well-known that Pocklington's integral equation [1] for a perfectly conducting wire of straight configuration shown in Fig.1.1 is given by

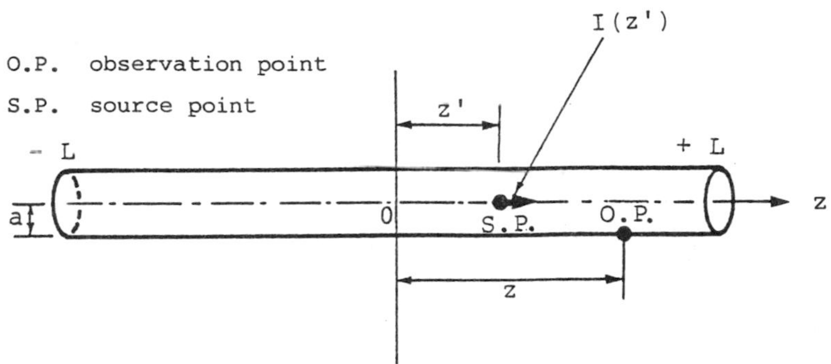

Fig.1.1 A single straight cylindrical antenna.

$$\frac{1}{j\omega\varepsilon}\int_{-L}^{+L} I(z')\left[\frac{\partial^2 V(z,z')}{\partial z^2} + \beta^2 V(z,z')\right] dz' + E_z^i(z) = 0 , \tag{1.1}$$

where I(z') is an equivalent filamentary line source, 2L is the wire length, β is the phase constant ($=\omega\sqrt{\mu\varepsilon}=2\pi/\lambda$, λ=wavelength), $E_z^i(z)$ is the incident field or impressed field, and V(z,z') is defined as

$$V(z,z') = \frac{1}{4\pi}\,\frac{e^{-j\beta r(z,z')}}{r(z,z')} , \tag{1.2}$$

where

$$r(z,z') = \sqrt{a^2 + (z-z')^2} . \tag{1.3}$$

Eqn.(1.1) is derived under the assumption that the current is distributed uniformly around the straight cylinder whose radius is "a", where $a \ll \lambda$, and that it flows only in the Z-axis direction.

The most common procedure of solving Pocklington's integral eqn.(1.1) is called the moment method [3], which reduces the integral equation to a set of simultaneous linear algebraic equations. The moment method is briefly summarised by taking two steps.

The first step is to approximate the current I(z') by a series of expansion functions J_n such that

$$I(z') = \sum_n I_n J_n(z') , \tag{1.4}$$

where I_n's are complex expansion coefficients. We assume the expansion functions as

$$J_n(z') = \begin{cases} J_n(z') & \text{for } z' \text{ in } \Delta z_n \\ 0 & \text{otherwise .} \end{cases} \tag{1.5}$$

Substituting eqn.(1.4) into eqn.(1.1), we calculate the value of the left side of eqn.(1.1), or the residual R(z),

$$R(z) = \sum_n I_n \int_{\Delta z_n} J_n(z') \, \Pi(z,z') \, dz' + E_z^i(z) , \tag{1.6}$$

where

$$\Pi(z,z') = \frac{1}{j\omega\varepsilon} \left[\frac{\partial^2 V(z,z')}{\partial z^2} + \beta^2 V(z,z') \right] . \tag{1.7}$$

The second step is to force the integration of weighted residuals to be zero, that is,

$$\int W_m(z)R(z)\,dz = 0 \,, \qquad (\, m = 1,2, \ldots\ldots \,) \qquad (1.8)$$

where $W_m(z)$ is called a weighting function or a testing function. When the weighting function is given by

$$W_m(z) = \begin{cases} W_m(z) & \text{for } z \text{ in } \Delta z_m \\ 0 & \text{otherwise ,} \end{cases} \qquad (1.9)$$

eqn.(1.8) becomes

$$\sum_n I_n \int_{\Delta z_m} W_m(z) \int_{\Delta z_n} J_n(z')\, \Pi(z,z')\, dz'dz$$
$$= - \int_{\Delta z_m} W_m(z) E_z^i(z)\, dz \,.$$
$$(\, m = 1,2, \ldots\ldots \,) \qquad (1.10)$$

Therefore, we have a set of simultaneous linear equations,

$$\sum_n Z_{mn}\, I_n = V_m \,, \qquad (\, m = 1,2, \ldots\ldots \,) \qquad (1.11)$$

where

$$Z_{mn} = \int_{\Delta z_m} W_m(z) \int_{\Delta z_n} J_n(z')\, \Pi(z,z')\, dz'dz \,, \qquad (1.12)$$

and

$$V_m = -\int_{\Delta z_m} W_m(z)\, E_z^i(z)\, dz \quad . \qquad (1.13)$$

Eqn.(1.11) can be written in matrix form $[Z_{mn}][I_n] = [V_m]$, and the solution is symbolically given by $[I_n] = [Z_{mn}]^{-1}[V_m]$. In practice, the unknown current vector $[I_n]$ is obtained by standard techniques.

1.3. PIECEWISE SINUSOIDAL GALERKIN METHOD

Once the expansion and weighting functions are chosen, eqn.(1.12) may be concretely calculated. Although there are many sets of expansion functions, we should choose one which closely resembles the anticipated form of the current on the wire. Fig.1.2 shows an example of a piecewise sinusoidal expansion function

$$J_n(z') = \begin{cases} \dfrac{\sin\beta(z' - z_{n-1})}{\sin\beta(z_n - z_{n-1})} & z_{n-1} \leq z' < z_n \\[2ex] \dfrac{\sin\beta(z_{n+1} - z')}{\sin\beta(z_{n+1} - z_n)} & z_n \leq z' < z_{n+1} \ . \end{cases}$$

(1.14)

When the weighting function is the same as the expansion function of eqn.(1.14), which is often referred to as a

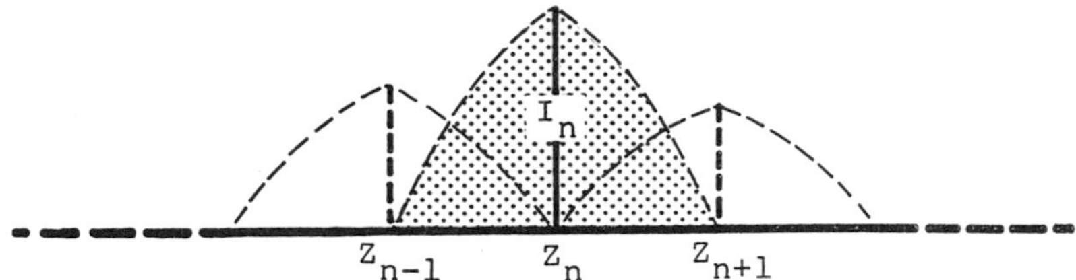

Fig.1.2 Piecewise sinusoidal expansion function.

piecewise sinusoidal Galerkin method, Z_{mn}'s in eqn.(1.12) become

$$Z_{mn} = \int_{z_{m-1}}^{z_m} \frac{\sin\beta(z - z_{m-1})}{\sin\beta(z_m - z_{m-1})} E_n(z)\, dz + \int_{z_m}^{z_{m+1}} \frac{\sin\beta(z_{m+1} - z)}{\sin\beta(z_{m+1} - z_m)} E_n(z)\, dz \,, \qquad (1.15)$$

where

$$E_n(z) = \int_{z_{n-1}}^{z_{n+1}} J_n(z')\, \Pi(z,z')\, dz'$$

$$= \frac{j30}{\sin\beta(z_n - z_{n-1})} \left[\frac{e^{-j\beta R_n^-}}{R_n^-} \cos\beta(z_n - z_{n-1}) - \frac{e^{-j\beta R_{n-1}^-}}{R_{n-1}^-} \right] + \frac{j30}{\sin\beta(z_{n+1} - z_n)} \left[\frac{e^{-j\beta R_n^-}}{R_n^-} \cos\beta(z_{n+1} - z_n) - \frac{e^{-j\beta R_n^+}}{R_n^+} \right] . \qquad (1.16)$$

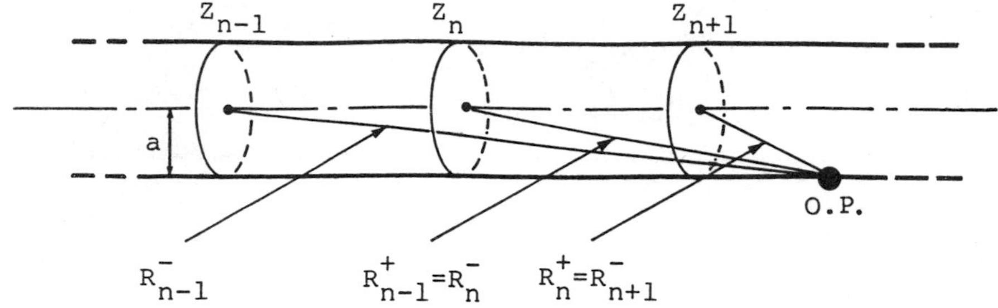

Fig.1.3 Co-ordinates for a straight wire antenna.

R_{n-1}^{-}, R_n^{-}, and R_n^{+} are shown in Fig.1.3, and are defined as

$$(R_{n-1}^{-})^2 = a^2 + (z - z_{n-1})^2 , \tag{1.17}$$

$$(R_n^{-})^2 = a^2 + (z - z_n)^2 , \tag{1.18}$$

$$(R_n^{+})^2 = a^2 + (z - z_{n+1})^2 . \tag{1.19}$$

Similarly, for a bent wire shown in Fig.1.4 we have Z_{mn}'s [5][10][11][12],

$$Z_{mn} = \int_{z_{m-1}}^{z_m} \frac{\sin\beta(z - z_{m-1})}{\sin\beta(z_m - z_{m-1})} \hat{z}_{m-1} \cdot \bar{e}_{n:wire} \, dz + \int_{z_m}^{z_{m+1}} \frac{\sin\beta(z_{m+1} - z)}{\sin\beta(z_{m+1} - z_m)} \hat{z}_m \cdot \bar{e}_{n:wire} \, dz \tag{1.20}$$

where

$$\bar{e}_{n:wire} = -j30 \left[\left(\frac{\bar{P}_{n-1}}{\rho_{n-1}\sin\beta e_{n-1}} + \frac{\bar{P}_n}{\rho_n \sin\beta e_n} \right) - \left(\frac{\bar{Z}_{n-1}}{\sin\beta e_{n-1}} + \frac{\bar{Z}_n}{\sin\beta e_n} \right) \right] . \quad (1.21)$$

$\bar{P}_{n-1}$, $\bar{P}_n$, $\bar{Z}_{n-1}$ and $\bar{Z}_n$ in eqn.(1.21) are defined as

$$\bar{P}_{n-1} = \left[\cos\beta e_{n-1} \cdot \cos\Theta^{+}_{n-1} \cdot e^{-j\beta R^{+}_{n-1}} - \cos\Theta^{-}_{n-1} \cdot e^{-j\beta R^{-}_{n-1}} - j\sin\beta e_{n-1} \cdot e^{-j\beta R^{+}_{n-1}} \right] \hat{\rho}_{n-1} , \quad (1.22)$$

$$\bar{P}_n = \left[\cos\beta e_n \cdot \cos\Theta^{-}_n \cdot e^{-j\beta R^{-}_n} - \cos\Theta^{+}_n \cdot e^{-j\beta R^{+}_n} + j\sin\beta e_n \cdot e^{-j\beta R^{-}_n} \right] \hat{\rho}_n , \quad (1.23)$$

$$\bar{Z}_{n-1} = \left[\cos\beta e_{n-1} \cdot \frac{e^{-j\beta R^{+}_{n-1}}}{R^{+}_{n-1}} - \frac{e^{-j\beta R^{-}_{n-1}}}{R^{-}_{n-1}} \right] \hat{z}_{n-1} , \quad (1.24)$$

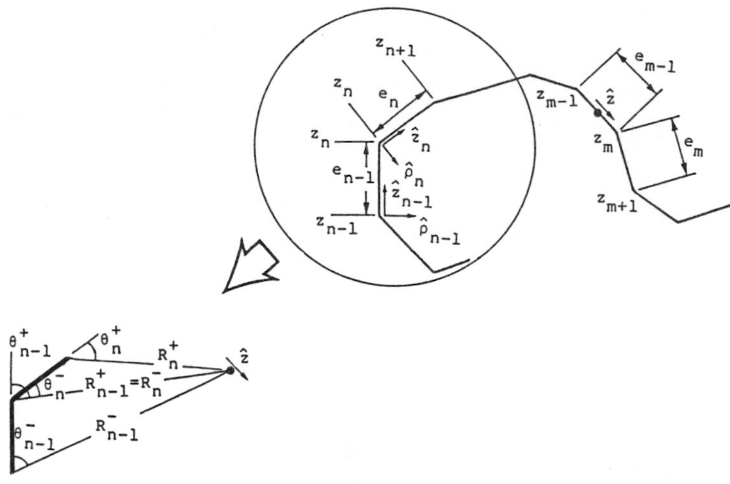

Fig.1.4 Co-ordinates for a bent wire antenna.

and

$$\bar{z}_n = \left[- \frac{e^{-j\beta R_n^+}}{R_n^+} + \cos\beta e_n \cdot \frac{e^{-j\beta R_n^-}}{R_n^-} \right] \hat{z}_n \, . \tag{1.25}$$

It is noted that in eqns.(1.22)-(1.25) the local cylindrical co-ordinates $(\phi_{n-1}, \rho_{n-1}, z_{n-1})$ and (ϕ_n, ρ_n, z_n) are used for the two wires n-1 and n, respectively. The unit vectors corresponding to these co-ordinates are symbolised as $(\hat{\phi}_{n-1}, \hat{\rho}_{n-1}, \hat{z}_{n-1})$ and $(\hat{\phi}_n, \hat{\rho}_n, \hat{z}_n)$, respectively. It is also noted that the angles $(\Theta_{n-1}^-, \Theta_{n-1}^+, \Theta_n^-, \Theta_n^+)$ and the lengths of the elements (e_{n-1}, e_n) are illustrated in Fig.1.4.

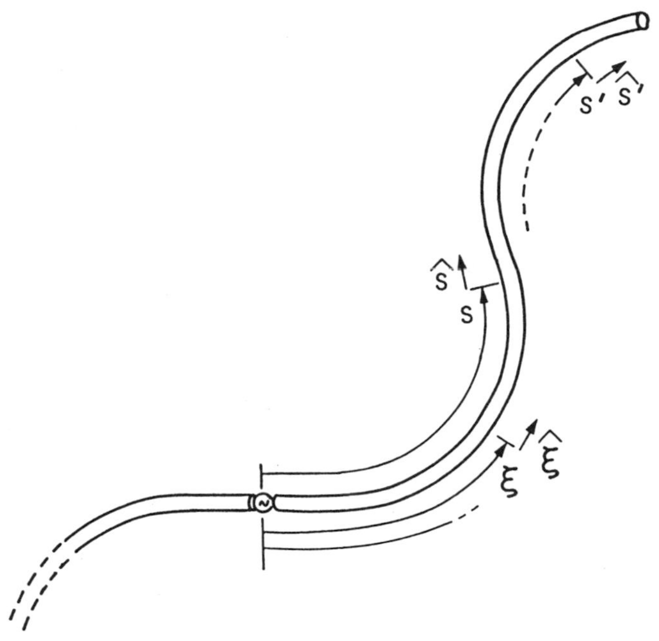

Fig.1.5 Thin wire antenna of arbitrary structure.

1.4. MEI'S INTEGRAL EQUATION AND POINT-MATCHING METHOD

Eqn.(1.26) is an integral equation derived by K.K.Mei [6] for thin wire antennas of arbitrary geometry shown in Fig.1.5:

$$\int_{\substack{\text{antenna}\\ \text{length}}} I(s')(\pi_1-\pi_2-\pi_3)ds'$$

$$= B\cos\beta s - \frac{jV_0}{2Z_0}\sin\beta|s| \;, \quad (1.26)$$

where

$$\pi_1 = G(s,s')\hat{s}\cdot\hat{s}' \ , \tag{1.27}$$

$$\pi_2 = \int_0^s G(\xi,s')\,\frac{d\hat{\xi}}{d\xi}\cdot\hat{s}' \cos\beta(s-\xi)\,d\xi \ , \tag{1.28}$$

$$\pi_3 = \int_0^s \left[\frac{\partial G(\xi,s')}{\partial\xi}(\hat{\xi}\cdot\hat{s}') + \frac{\partial G(\xi,s')}{\partial s'} \right] \cos\beta(s-\xi)\,d\xi \ . \tag{1.29}$$

The symbols in eqns.(1.26) — (1.29) are defined as follows : B is the constant of integration which must be consistent with the boundary condition that the current is zero at the wire end; V_0 and Z_0 are, respectively, the voltage applied and the intrinsic impedance of free space; $\beta = 2\pi/\lambda$ (λ = free-space wavelength); s, s', and ξ are the distances measured from the feed gap along the arm; $\hat{s}$, $\hat{s}'$, and $\hat{\xi}$ are tangential unit vectors at s, s', and ξ, respectively; G(s,s') is the free-space Green's function defined as

$$G(s,s') = \frac{1}{4\pi}\,\frac{e^{-j\beta r(s,s')}}{r(s,s')} \ , \tag{1.30}$$

in which r(s,s') is the distance between an observation point of s and a source point of s'(notice : eqn.(1.30) becomes eqn.(1.2), when antenna geometry is linear).

We assume that the current I(s') can be approximated by $\sum I_n J_n(s')$ as shown in eqn.(1.4), and transform eqn.(1.26) into

$$\sum_n I_n \int_{\substack{\text{antenna}\\ \text{length}}} J_n(s')\ \pi(s,s')\ ds' = f(s)\ , \tag{1.31}$$

where $\pi(s,s') = \pi_1 - \pi_2 - \pi_3$, and f(s) represents the right side of eqn.(1.26). Enforcing eqn.(1.31) at N different points in the antenna arm, we obtain N linear equations

$$\sum_n I_n \int_{\substack{\text{antenna}\\ \text{length}}} J_n(s')\ \pi(s_m,s')\ ds' = f(s_m), \qquad m = 1,2,\cdots,N \tag{1.32}$$

The process of obtaining eqn.(1.32) is called a point-matching method.

If we choose $J_n(s')$ as

$$J_n(s') = \begin{cases} 1 & \text{for } s' \text{ in } \Delta s_n \\ 0 & \text{otherwise ,} \end{cases} \tag{1.33}$$

then eqn.(1.32) becomes

$$\sum_n I_n \int_{\Delta s_n} \pi(s_m,s')\ ds' = f(s_m)\ , \qquad m = 1,2,\cdots,N \tag{1.34}$$

which corresponds to eqn.(1.11) and may be solved for I_n by standard techniques. Comments on the choice of $J_n(s')$ other than eqn.(1.33) are found in Reference [13].

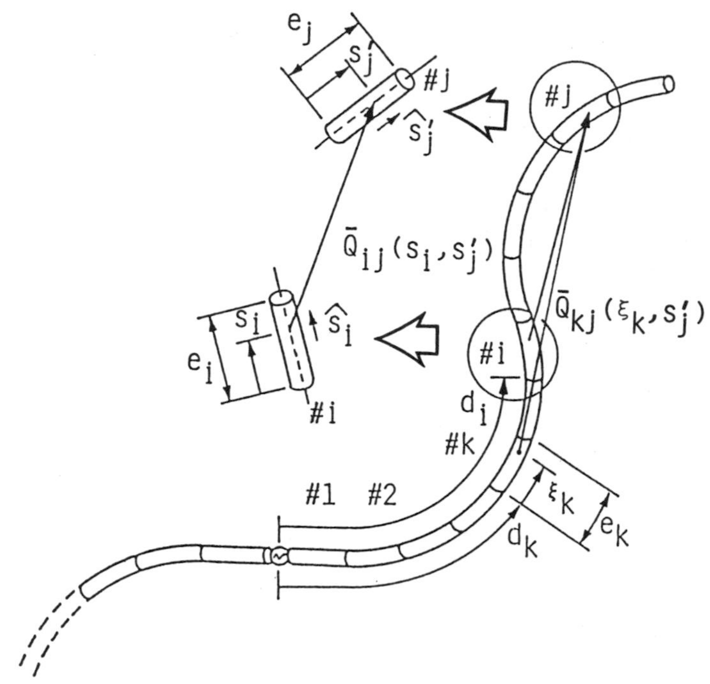

Fig.1.6 Co-ordinate system of thin wire antenna of arbitrary structure.

1.5. NAKANO'S INTEGRAL EQUATION

The kernel, $\pi_1 - \pi_2 - \pi_3$, in eqn.(1.26) contains integrals and derivatives, as seen from eqns.(1.28) and (1.29). If we subdivide an antenna arm into some elements which can be regarded as being linear, another integral equation whose kernel contains neither derivatives nor integrals is obtained. This integral equation is called Nakano's integral equation [7][8] and will be given by eqn.(1.63) with the kernel of eqn.(1.62) on page 21.

There are two ways to derive Nakano's integral equation. One is to make use of Mei's integral equation (1.26) and simplify its kernel into a closed form which contains neither derivatives nor integrals. The other is to make use of Hallen's equation for a single straight wire and to apply the recurrence formulas concerning the continuity of vector potential and continuity of scalar potential at the bend points of the elements of subdivided antenna arm [14][15]. In this section, the former derivation is described in detail.

1.5.1. Closed form of π_2

After the antenna arm is subdivided, s_i, s_j, and ξ_k are defined as $s = d_i + s_i$, $s' = d_j + s_j'$, and $\xi = d_k + \xi_k$, where d_i, d_j, and d_k are the distances along the antenna arm from the feed gap to the starting points of the i-th, the j-th, and the k-th elements, respectively, as shown in Fig.1.6. We give the directions of these elements by using tangential unit vectors $\hat{s}_i$, $\hat{s}_j'$, and $\hat{\xi}_k$ (= $\hat{s}_k$), respectively, and write a vector $\bar{Q}_{kj}(\xi_k,s_j')$ which extends from ξ_k on the k-th element axis to s_j' on the j-element axis as

$$\bar{Q}_{kj}(\xi_k,s_j') = -\xi_k\hat{s}_k + \bar{Q}_{kj}(0,0) + s_j'\hat{s}_j' \;. \tag{1.35}$$

The subdivision of antenna arm readily reduces π_2 in eqn.(1.28) to a closed form

$$\pi_2 = \sum_{k=1}^{i-1} G_{kj}(e_k,s_j')(\hat{s}_{k+1} - \hat{s}_k)\cdot\hat{s}_j' \cos \beta D_{ik}(s_i,e_k) \tag{1.36}$$

where e_k is the length of the k-th element. $G_{kj}(e_k,s_j')$ and $D_{ik}(s_i,e_k)$ in eqn.(1.36) obey the following definitions

$$G_{kj}(e_k,s_j')=\frac{e^{-j\beta r_{kj}(e_k,s_j')}}{4\pi r_{kj}(e_k,s_j')}\ , \tag{1.37}$$

$$D_{ik}(s_i,e_k)= d_i + s_i - d_k - e_k\ , \tag{1.38}$$

where $r_{kj}(e_k,s_j')$ in eqn.(1.37) is defined as

$$r_{kj}(e_k,s_j')= [\,|\bar{Q}_{kj}(e_k,s_j')|^2 + a^2\,]^{\frac{1}{2}}\ , \tag{1.39}$$

where "a" is the radius of the antenna wire.

1.5.2. Closed form of π_3

We focus our attention on simplifying π_3. Eqn.(1.29) can be separated into two groups after the subdivision of the antenna arm

$$\pi_3= (\sum_{k=1}^{i-1} \gamma_k) + \gamma_i\ , \tag{1.40}$$

where

$$\gamma_k=\int_0^{e_k} [\ \frac{\partial G_{kj}(\xi_k,s_j')}{\partial \xi_k}\ (\hat{s}_k\cdot\hat{s}_j') + \frac{\partial G_{kj}(\xi_k,s_j')}{\partial s_j'}\ ]\cos \beta D_{ik}(s_i,\xi_k)d\xi_k \tag{1.41}$$

and

$$\gamma_i = \int_0^{s_i} \left[\frac{\partial G_{ij}(\xi_i, s_j')}{\partial \xi_i} (\hat{s}_i \cdot \hat{s}_j') + \frac{\partial G_{ij}(\xi_i, s_j')}{\partial s_j'} \right] \cos \beta D_{ii}(s_i, \xi_i)\, d\xi_i \; . \tag{1.42}$$

Since γ_k and γ_i have the same form, we proceed to simplify only γ_k. Using integration of parts, we transform eqn.(1.41) to

$$\gamma_k = (\hat{s}_k \cdot \hat{s}_j') \, [G_{kj}(e_k, s_j') \cos \beta D_{ik}(s_i, e_k) - G_{kj}(0, s_j') \cos \beta D_{ik}(s_i, 0)] + \frac{\partial}{\partial s_j'} \int_0^{e_k} G_{kj}(\xi_k, s_j') \cos \beta D_{ik}(s_i, \xi_k)\, d\xi_k - (\hat{s}_k \cdot \hat{s}_j') \, \beta \int_0^{e_k} G_{kj}(\xi_k, s_j') \sin \beta D_{ik}(s_i, \xi_k)\, d\xi_k \tag{1.43}$$

In order to calculate the last two terms involved in integration, we introduce

$$m_{kj}^{+} = r_{kj}(\xi_k, s_j') + \bar{Q}_{kj}(\xi_k, s_j') \cdot \hat{s}_k \tag{1.44}$$

and

$$m_{kj}^{-} = r_{kj}(\xi_k, s_j') - \bar{Q}_{kj}(\xi_k, s_j') \cdot \hat{s}_k \; . \tag{1.45}$$

Using eqns.(A-1) and (A-2) in Appendix-I, we may write eqn.(1.43) in the following form:

$$\gamma_k = (\hat{s}_k \cdot \hat{s}_j')[\, G_{kj}(e_k, s_j') \cos \beta D_{ik}(s_i, e_k)$$

$$- G_{kj}(0, s_j') \cos \beta D_{ik}(s_i, 0)$$

$$+ \frac{1}{4\pi} \frac{\partial}{\partial s_j'} (\psi^- - \psi^+)$$

$$+ j\frac{\beta}{4\pi} (\hat{s}_k \cdot \hat{s}_j')[\psi^- + \psi^+] \,, \qquad (1.46)$$

where

$$\psi^+ = \frac{1}{2} e^{j\beta C_{kj}} \int_{U_{kj}^+(0, s_j')}^{U_{kj}^+(e_k, s_j')} \frac{e^{-j\beta m_{kj}^+}}{m_{kj}^+} dm^+{}_{kj} \qquad (1.47)$$

and

$$\psi^- = \frac{1}{2} e^{-j\beta C_{kj}} \int_{U_{kj}^-(0, s_j')}^{U_{kj}^-(e_k, s_j')} \frac{e^{-j\beta m_{kj}^-}}{m_{kj}^-} dm^-{}_{kj} \,. \qquad (1.48)$$

C_{kj} and $U_{kj}^{\pm}$ in eqns.(1.47) and (1.48) are defined as

$$C_{kj} = \bar{Q}_{kj}(0, s_j') \cdot \hat{s}_k - D_{ik}(s_i, 0) \,, \qquad (1.49)$$

$$U^{+}_{kj}(e_k,s'_j)= r_{kj}(e_k,s'_j) + \bar{Q}_{kj}(e_k,s'_j)\cdot\hat{s}_k \ , \tag{1.50}$$

$$U^{-}_{kj}(e_k,s'_j)= r_{kj}(e_k,s'_j) - \bar{Q}_{kj}(e_k,s'_j)\cdot\hat{s}_k \ , \tag{1.51}$$

$$U^{+}_{kj}(0,s'_j)= r_{kj}(0,s'_j) + \bar{Q}_{kj}(0,s'_j)\cdot\hat{s}_k \ , \tag{1.52}$$

$$U^{-}_{kj}(0,s'_j)= r_{kj}(0,s'_j) - \bar{Q}_{kj}(0,s'_j)\cdot\hat{s}_k \ . \tag{1.53}$$

After differentiating ψ^+ and ψ^- with respect to s'_j in eqn.(1.46) we have [see Appendix-II]

$$\gamma_k= \eta_{kij}(e_k,s_i,s'_j) - \eta_{kij}(0,s_i,s'_j) \ , \tag{1.54}$$

where

$$\eta_{kij}(e_k,s_i,s'_j)=[\bar{Q}_{kj}(e_k,s'_j)\cdot\hat{s}_k G_{kj}(e_k,s'_j) \cos \beta D_{ik}(s_i,e_k) + j\frac{1}{4\pi} e^{-j\beta r_{kj}(e_k,s'_j)} \sin \beta D_{ik}(s_i,e_k)] \, g_{kj} \tag{1.55}$$

and

$$\eta_{kij}(0,s_i,s'_j)= [\bar{Q}_{kj}(0,s'_j)\cdot\hat{s}_k \, G_{kj}(0,s'_j) \cos \beta D_{ik}(s_i,0) + j\frac{1}{4\pi} e^{-j\beta r_{kj}(0,s'_j)} \sin \beta D_{ik}(s_i,0)] \, g_{kj} \ . \tag{1.56}$$

$g_{kj}(0,s_j')$ in eqns.(1.55) and (1.56) is defined as

$$g_{kj}=\frac{\bar{Q}_{kj}(0,s_j')\cdot\hat{s}_j' - [\bar{Q}_{kj}(0,s_j')\cdot\hat{s}_k](\hat{s}_k\cdot\hat{s}_j')}{|\bar{Q}_{kj}(0,s_j')|^2 - [\bar{Q}_{kj}(0,s_j')\cdot\hat{s}_k]^2 + a^2} \quad . \tag{1.57}$$

Similarly, γ_i is given by

$$\gamma_i = \eta_{iij}(s_i,s_i,s_j') - \eta_{iij}(0,s_i,s_j') \quad , \tag{1.58}$$

where $\eta_{iij}(s_i,s_i,s_j')$ and $\eta_{iij}(0,s_i,s_j')$ may be calculated by using eqns.(1.55) and (1.56), respectively.

$$\eta_{iij}(s_i,s_i,s_j') = [\bar{Q}_{ij}(s_i,s_j')\cdot\hat{s}_i \; G_{ij}(s_i,s_j')] \; g_{ij} \tag{1.59}$$

and

$$\eta_{iij}(0,s_i,s_j') = [\bar{Q}_{ij}(0,s_j')\cdot\hat{s}_i \; G_{ij}(0,s_j') \cos\beta s_i + j\frac{1}{4\pi} e^{-j\beta r_{ij}(0,s_j')} \sin\beta s_i] \; g_{ij} \quad . \tag{1.60}$$

Substituting eqns.(1.54) and (1.58) into eqn.(1.40), we obtain

$$\pi_3 = \eta_{iij}(s_i,s_i,s_j') - \eta_{iij}(0,s_i,s_j') + \sum_{k=1}^{i-1} [\eta_{kij}(e_k,s_i,s_j') - \eta_{kij}(0,s_i,s_j')] \quad . \tag{1.61}$$

1.5.3. Final form of kernel

From eqns.(1.27), (1.36) and (1.61) the kernel $\pi_1 - \pi_2 - \pi_3 \equiv \pi_{ij}(s_i,s_j')$ becomes

$$\pi_{ij}(s_i,s_j') = G_{ij}(s_i,s_j')\hat{s}_i \cdot \hat{s}_j'$$

$$- \eta_{iij}(s_i,s_i,s_j') + \eta_{iij}(0,s_i,s_j')$$

$$- \sum_{k=1}^{i-1} [G_{kj}(e_k,s_j')(\hat{s}_{k+1} - \hat{s}_k)\cdot\hat{s}_j' \cos \beta D_{ik}(s_i,e_k)$$

$$+ \eta_{kij}(e_k,s_i,s_j') - \eta_{kij}(0,s_i,s_j')] \qquad (1.62)$$

It should be noted that the kernel $\pi_{ij}(s_i,s_j')$ is expressed in a simple algebraic form, containing neither derivatives nor integrals.

With the kernel of eqn.(1.62), we transform eqn.(1.26) into

$$\sum_j \int_{\substack{j\text{-th}\\ \text{element}}} I_j(s_j')\pi_{ij}(s_i,s_j')ds_j'$$

$$= B \cos \beta(d_i + s_i) - \frac{jV_0}{2Z_0} \sin \beta|d_i + s_i| . \qquad (1.63)$$

A fundamental task for programming this equation requires only co-ordinates of starting and ending points of subdivided elements of an antenna arm. This leads to an advantage that we can easily handle antennas of arbitrary geometry with less computational time.

1.6. A PAIR OF INTEGRAL EQUATIONS FOR A SYSTEM COMPOSED OF WIRES AND SLOTS

So far, we have referred to a single arbitrary thin wire and shown how to derive Nakano's equation with a closed kernel. It is noted that the closed kernel is also found in problems of the electromagnetic coupling between thin wires and narrow slots shown in Fig.1.7 [9][16][17]. The assumptions taken in this antenna problem are as follows: the N wires are perfectly conducting and are thin relative to the wavelength of the incident electromagnetic field ($\bar{E}^{in}$, $\bar{H}^{in}$); the M narrow slots behind the N wires are in a perfectly conducting, vanishingly thin screen of infinite extent.

Unknown electric current $I_{nj}(s'_{nj})$ and magnetic current $m_{nj}(s'_{nj})$, where s'_{nj} is the distance measured along the nj-th wire (slot) element from its starting point, are determined by solving a pair of integral equations:

$$jZ_0 \sum_{n=1}^{2N} \sum_{j=1}^{N_n} \int_0^{e_{nj}} I_{nj}(s'_{nj})\, \Pi_{mi\ nj}(s_{mi}, s'_{nj})\, ds'_{nj}$$

$$+ 2 \sum_{n=2N+1}^{2N+M} \sum_{j=1}^{N_n} \int_0^{e_{nj}} m_{nj}(s'_{nj})\, \Pi^{\#}_{mi\ nj}(s_{mi}, s'_{nj})\, ds'_{nj}$$

$$= C_m \cos\beta(d_{mi}+s_{mi}) + B_m \sin\beta(d_{mi}+s_{mi})$$

for s_{mi} on the wires, (1.64)

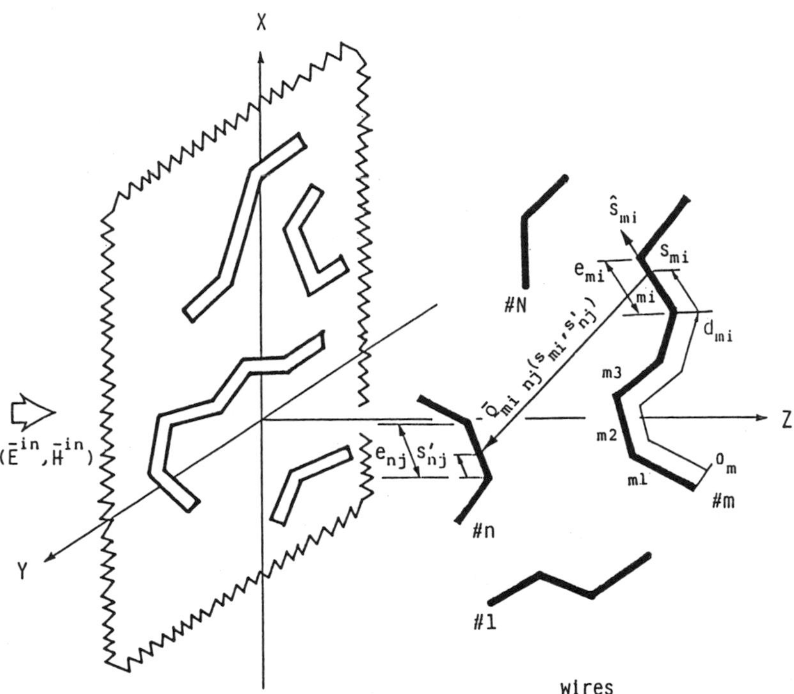

Fig.1.7(a) Co-ordinate of a system composed of N wires.

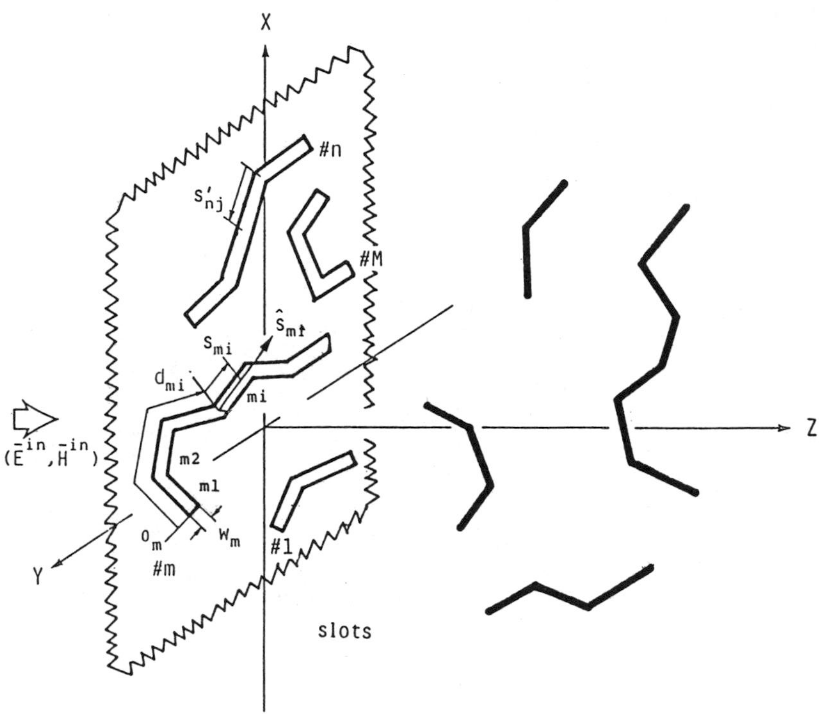

Fig.1.7(b) Co-ordinate of a system composed of M slots.

and

$$-\frac{1}{2}\sum_{n=1}^{2N}\sum_{j=1}^{N_n}\int_0^{e_{nj}} I_{nj}(s'_{nj})\Pi^{\#}_{mi\ nj}(s_{mi},s'_{nj})\,ds'_{nj}$$

$$+\,j\frac{1}{Z_0}\sum_{n=2N+1}^{2N+M}\sum_{j=1}^{N_n}\int_0^{e_{nj}} 2m_{nj}(s'_{nj})\Pi_{mi\ nj}(s_{mi},s'_{nj})\,ds'_{nj}$$

$$= C^{\#}_m \cos\beta(d_{mi}+s_{mi}) + B^{\#}_m \sin\beta(d_{mi}+s_{mi})$$

$$+\int_0^{d_{mi}+s_{mi}} \hat{s}_{mi}\cdot\bar{H}^{in}(\eta_m)\sin\beta(d_{mi}+s_{mi}-\eta_m)\,d\eta_m$$

<u>for s_{mi} on the slots,</u> (1.65)

where the symbols are defined as follows:

N_n : number of wire (slot) elements (= number of junction points + 1),

e_{nj} : length of the nj-th wire (slot) element,

s_{mi} : distance measured along the mi-th wire (slot) element from its starting point,

s'_{nj} : distance measured along the nj-th wire (slot) element from its starting point,

d_{mi} : distance measured along the m-th wire (slot) from its origin to the starting point of the mi-th wire (slot) element,

Z_0 : intrinsic impedance of free space (= 120π),

C_m and B_m : unknown coefficients which are determined by using the boundary condition on the electric current at the ends of the m-th wire,

$C_m^{\#}$ and $B_m^{\#}$: unknown coefficients which are determined by using the boundary condition on the magnetic current at the ends of the m-th slots,

$\hat{s}_{mi}$: unit vector tangential to the mi-th element.

The kernels in eqns.(1.64) and (1.65) are

$$\begin{aligned}
\Pi_{mi\ nj}(s_{mi},s'_{nj}) &\equiv \Pi_{IJ}(s_I,s'_J) \\
&= V_{IJ}(s_I,s'_J)\hat{s}_I\cdot\hat{s}'_J \\
&- [\ \eta_{IIJ}(s_I,s_I,s'_J) - \eta_{IIJ}(0,s_I,s'_J)\ ] \\
&- \sum_{k=1}^{i-1}\ [\ V_{KJ}(e_K,s'_J)(\hat{s}_{K+1}-\hat{s}_K)\cdot\hat{s}'_J\cos\beta D_{IK}(s_I,e_K) \\
&+ \eta_{KIJ}(e_K,s_I,s'_J) - \eta_{KIJ}(0,s_I,s'_J)\ ], \qquad (1.66)
\end{aligned}$$

and

$$\Pi^{\#}_{mi\ nj}(s_{mi},s'_{nj}) \equiv \Pi^{\#}_{IJ}(s_I,s'_J)$$

$$= - [\ \eta^{\#}_{IIJ}(s_I,s_I,s'_J) - \eta^{\#}_{IIJ}(0,s_I,s'_J)\ ]$$

$$- \sum_{k=1}^{i-1} [\ \eta^{\#}_{KIJ}(e_K,s_I,s'_J) - \eta^{\#}_{KIJ}(0,s_I,s'_J)\ ],$$

(1.67)

where subscripts of mi, nj, mk, and m<u>k+1</u> are, for convenience, replaced by capitals of I, J, K, and K+1,respectively (notice : summation of $\sum$ is carried out with respect to a small letter of k). Green's function V_{IJ}, distance D_{IK}, η_{KIJ}, and $\eta^{\#}_{KIJ}$ obey the definitions of eqns.(A-10), (A-12), (A-16), and (A-17) in Appendix-III, respectively. It should be noted that the kernels Π_{IJ} of eqn.(1.66) and $\Pi^{\#}_{IJ}$ of eqn.(1.67) contain neither derivatives nor integrals, and that Π_{IJ} of eqn.(1.66) is basically the same as eqn.(1.62) except for the expressions on the subscripts.

The scalar potentials for the wires and slots are given by

$$\phi_{mi}(s_{mi}) = - C_m \sin\beta(d_{mi}+s_{mi}) + B_m \cos\beta(d_{mi}+s_{mi})$$

$$- jZ_0 \sum_{n=1}^{2N} \sum_{j=1}^{N_n} \int_0^{e_{nj}} I_{nj}(s'_{nj})\Phi^{\#}_{mi\ nj}(s_{mi},s'_{nj})ds'_{nj}$$

$$+ 2 \sum_{n=2N+1}^{2N+M} \sum_{j=1}^{N_n} \int_0^{e_{nj}} m_{nj}(s'_{nj})\Phi_{mi\ nj}(s_{mi},s'_{nj})ds'_{nj}$$

(1.68)

<u>for s_{mi} on the wires,</u>

and

$$\phi^{\#}_{mi}(s_{mi}) = - C^{\#}_{m}\sin\beta(d_{mi}+s_{mi}) + B^{\#}_{m}\cos\beta(d_{mi}+s_{mi})$$

$$+ \int_{0}^{d_{mi}+s_{mi}} \hat{s}_{mi}\cdot\bar{H}^{in}(\eta_m)\cos\beta(d_{mi}+s_{mi}-\eta_m)\, d\eta_m$$

$$- \frac{1}{2} \sum_{n=1}^{2N} \sum_{j=1}^{N_n} \int_{0}^{e_{nj}} I_{nj}(s'_{nj})\Phi_{mi\ nj}(s_{mi},s'_{nj})\, ds'_{nj}$$

$$- j\frac{1}{Z_0} \sum_{n=2N+1}^{2N+M} \sum_{j=1}^{N_n} \int_{0}^{e_{nj}} 2m_{nj}(s'_{nj})\Phi^{\#}_{mi\ nj}(s_{mi},s'_{nj})ds'_{nj}$$

(1.69)

<u>for s_{mi} on the slots,</u>

where

$$\Phi^{\#}_{mi\ nj}(s_{mi},s'_{nj}) \equiv \Phi^{\#}_{IJ}(s_I,s'_J)$$

$$= \xi^{\#}_{IIJ}(s_I,s_I,s'_J) - \xi^{\#}_{IIJ}(0,s_I,s'_J)$$

$$+ \sum_{k=1}^{i-1} [\xi^{\#}_{KIJ}(e_K,s_I,s'_J) - \xi^{\#}_{KIJ}(0,s_I,s'_J)]$$

$$+ \sum_{k=1}^{i-1} [V_{KJ}(e_K,s'_J)[\hat{s}_{K+1}-\hat{s}_K]\cdot\hat{s}'_J\sin\beta D_{IK}(s_I,e_K)]$$

(1.70)

and

$$\Phi_{mi\,nj}(s_{mi},s'_{nj}) \equiv \Phi_{IJ}(s_I,s'_J)$$

$$= \xi_{IIJ}(s_I,s_I,s'_J) - \xi_{IIJ}(0,s_I,s'_J)$$

$$+ \sum_{k=1}^{i-1} [\xi_{KIJ}(e_K,s_I,s'_J) - \xi_{KIJ}(0,s_I,s'_J)],$$

(1.71)

in which ξ_{KIJ} and $\xi^{\#}_{KIJ}$ obey the definitions of eqns.(A-16) and (A-17), respectively.

Eqns.(1.68) and (1.69) show that the scalar potentials are characterised by the kernels Φ and $\Phi^{\#}$, which have a closed form. It is noted that the kernel Φ associated with the magnetic current in eqn.(1.68) appears in the term associated with the electric current in eqn.(1.69), while the kernel $\Phi^{\#}$ associated with the electric current in eqn.(1.68) appears in the term associated with the magnetic current in eqn.(1.69). This property contributes to easier computer programming.

Fig.1.8 shows the numerical results of a system consisting of square slot and wire elements of one-wavelength periphery. The slot is illuminated by a linearly polarised wave and a circularly polarised wave. As expected, we find standing wave distributions and travelling wave distributions in Figs. (a) and (b), respectively.

The solutions for other situations are also available in References [16] and [18]. It should be noted that in Reference [18] spiral slot antennas fed by a single wire are analysed by different expressions [19] corresponding to Eqs.(1.64) and (1.65).

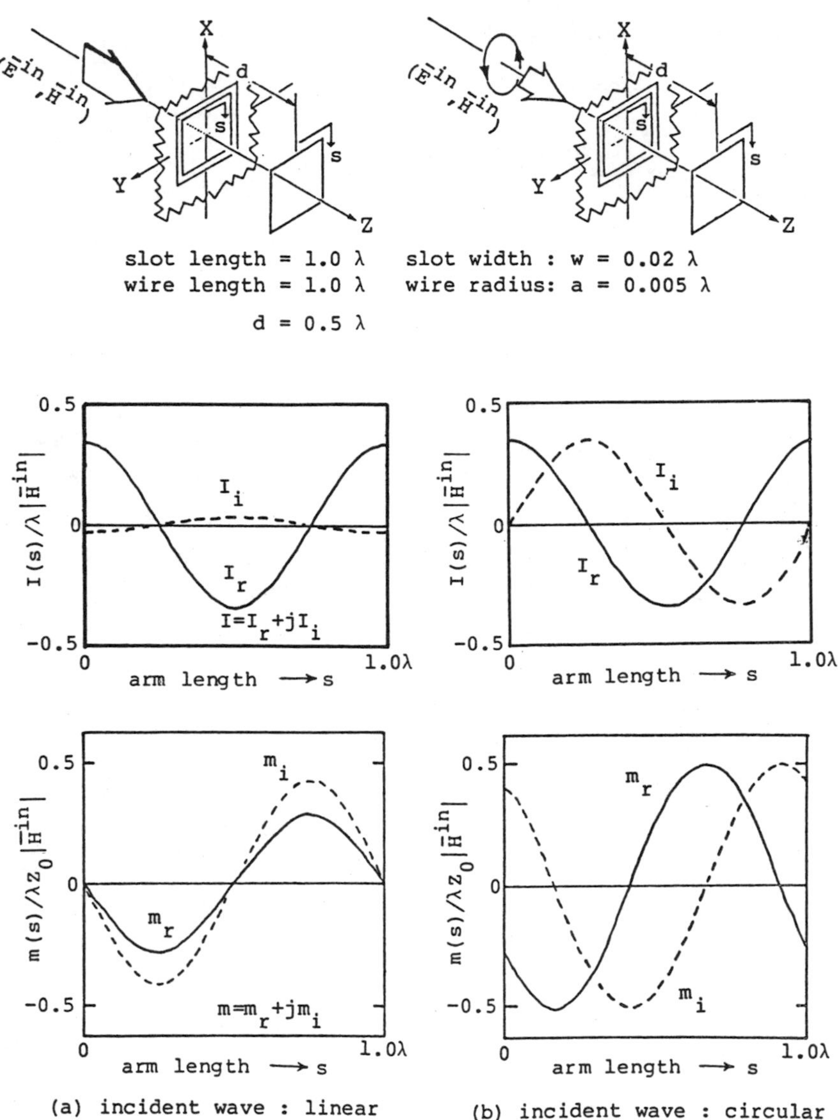

Fig.1.8 Electric current I(s) on a wire and magnetic current m(s) on a slot.

1.7. COMMENTS ON EVALUATION OF ANTENNA CHARACTERISTICS

Once the current distribution I(s') is determined, it is a familiar task to evaluate the electric field ($\bar{E} = E_\theta \hat{\theta} + E_\phi \hat{\phi}$) at a far-field point (spherical co-ordinates: R, θ, ϕ). Using two components of the far-field, E_θ and E_ϕ, we may calculate the axial ratio AR.

AR =

$$20 \log_{10} \left[\frac{|E_\phi|^2 \sin^2\tau + |E_\theta|^2 \cos^2\tau + |E_\phi||E_\theta| \cos\delta \sin 2\tau}{|E_\phi|^2 \cos^2\tau + |E_\theta|^2 \sin^2\tau - |E_\phi||E_\theta| \cos\delta \sin 2\tau} \right]^{1/2} \quad \text{[dB]}, \qquad (1.72)$$

where

$$2\tau = \tan^{-1} \frac{2|E_\phi||E_\theta| \cos \delta}{|E_\theta|^2 - |E_\phi|^2} \,. \qquad (1.73)$$

in which δ is the phase difference between the field components E_θ and E_ϕ.

The input impedance of an antenna, $Z_{in} = R_{in} + jX_{in}$, is determined by the ratio of the source voltage to the current I(0) at the input terminal. The power gain is given by

$$G = 10 \log_{10} \frac{[\, |E_\theta|^2 + |E_\phi|^2 \,]\, R^2}{30\, |I(0)|^2\, R_{in}} \quad \text{[dB]}. \qquad (1.74)$$

REFERENCES TO CHAPTER 1

[1] Pocklington, H.E., "Electrical oscillation in wires", Cambridge Phil. Soc. Proc., Vol.9, 1897, pp.324-332.

[2] Hallen, E., "Theoretical investigation into the transmitting and receiving qualities of antennae", Nova Acta Soc. Sci. Upsal., 1938, pp.1-4.

[3] Harrington, R.F., "Field computation by moment methods", (Macmillan, New York, 1968).

[4] Stutzman, W.L. and Thiele, G.A., "Antenna theory and design", (John Wiley and Sons, New York, 1981), Chapter 7.

[5] Richmond, J.H., "Radiation and scattering by thin-wire structures in the complex frequency domain", NASA CR-2396, 1974.

[6] Mei, K.K., "On the integral equations of thin wire antennas", IEEE Trans., AP-13, 1965, pp.374-378.

[7] Nakano, H., "The simplified expression for the kernel of Mei's integral equation", IECE TGAP, AP79-16, 1979.

[8] Nakano, H., "The integral equation for a system composed of many arbitrarily bent wires", IECE Trans., Vol.E65, No.6, 1982, pp.303-309.

[9] Nakano, H., Yamauchi, J., Eda, M. and Iwasaki, T., "Numerical analysis of electromagnetic couplings between wires and slots using integral equations", Proc.IEE Fourth International Conference on Antennas and Propagation, Coventry, 1985, pp.438-442.

[10] Tsukiji, T. and Yamasaki, M., "Analysis of broad band triangular double loop antenna", IECE Trans., Vol.J65-B, No.6, 1982, pp.769-776.

[11] Hirasawa, K., and Fujimoto, K., "On wire-grid method for analysis of wire antennas near/on a rectangular conducting body", IECE Trans., Vol.J65-B,No.4, 1982, pp.382-389.

[12] Nakano, H., Yamauchi, J., Yoshizawa, A., "The moment method for electromagnetic coupling between arbitrarily bent wire and slot structures", Proc. of 1985 IECE International Symposium on Antennas and Propagation, Vol.3, 222-5, 1985, pp.851-854.

[13] Yeh, Y.S. and Mei, K.K. "Theory of conical equiangular-spiral antennas : part 1 - numerical technique", IEEE Trans., AP-15, 1967, pp.634-639.

[14] Nakano, H., "Simplified integral equation on electromagnetic coupling between arbitrarily bent wires and slots", IEEE International Symposium on Antennas and Propagation Vol.1, APS-12-7, 1984, pp.413-416.

[15] Nakano, H., "Integral equations on electromagnetic coupling to wires through apertures(III)", IECE TGAP, AP83-60, 1983.

[16] Nakano, H., Harrington, R.F., "Integral equations on electromagnetic coupling to a wire through an aperture", IECE Trans., Vol.E66, No.6, 1983, pp.383-389.

[17] Nakano, H., Yamane, T., Yamauchi, J., "Hallen-type integral equation for a system composed of wires and slots", IECE Trans., Vol.E67, No.9, 1984, pp.516-522.

[18] Nakano, H., Tanabe, M., Yamauchi, J., Shafai, L., "Spiral slot antenna", Proc.IEE Fifth International Conference on Antennas and Propagation, York, 1987, pp.86-89.

[19] Nakano, H., Yamauchi, J., Yoshizawa, A., "The moment method for electromagnetic coupling between arbitrarily bent wire and slot structures", Proceedings of IECE International Conference on Antennas and Propagation, Kyoto, Japan, 1985, pp.851-854.

Appendix-I

Differentiating eqns.(1.44) and (1.45) with respect to ξ_k, we obtain

$$\frac{dm^+_{kj}}{d\xi_k} = \frac{-\, m^+_{kj}}{r_{kj}(\xi_k, s'_j)} \tag{A-1}$$

and

$$\frac{dm^-_{kj}}{d\xi_k} = \frac{m^-_{kj}}{r_{kj}(\xi_k, s'_j)} \, . \tag{A-2}$$

Appendix-II

We define H_k as

$$4\pi H_k = \frac{\partial}{\partial s'_j} [\psi^- - \psi^+] + j\beta(\hat{s}_k \cdot \hat{s}'_j)[\psi^- + \psi^+] \, . \tag{A-3}$$

The following relation is used for differentiations ψ^+ and ψ^- with respect to s'_j

$$\frac{\partial C_{kj}}{\partial s'_j} = \frac{\partial \bar{Q}_{kj}(0, s'_j)}{\partial s'_j} \cdot \hat{s}_k = \hat{s}'_j \cdot \hat{s}_k \, . \tag{A-4}$$

Thus, we obtain

$$4\pi H_k = h_k(e_k) - h_k(0) \, , \tag{A-5}$$

where

$$h_k(e_k) = e^{-j\beta r_{kj}(e_k,s_j')}\left[\frac{\cos \beta D_{ik}(s_i,e_k)}{r_{kj}(e_k,s_j')}\right.$$

$$\cdot \{\bar{Q}_{kj}(e_k,s_j')\cdot\hat{s}_k\, g_{kj} - \hat{s}_k\cdot\hat{s}_j'\}$$

$$\left.+ jg_{kj} \sin \beta D_{ik}(s_i,e_k)\right] , \qquad (A\text{-}6)$$

$$h_k(0) = e^{-j\beta r_{kj}(0,s_j')}\left[\frac{\cos \beta D_{ik}(s_i,0)}{r_{kj}(0,s_j')}\right.$$

$$\cdot \{\bar{Q}_{kj}(0,s_j')\cdot\hat{s}_k\, g_{kj} - \hat{s}_k\cdot\hat{s}_j'\}$$

$$\left.+ jg_{kj} \sin \beta D_{ik}(s_i,0)\right] , \qquad (A\text{-}7)$$

in which g_{kj} is defined as in eqn.(1.57).

Substituting (A-5) into eqn.(1.46), we obtain eqn.(1.54).

Appendix-III

(A) Symbols

$a_{mi} \equiv a_I$: wire radius of the mi-th wire or equivalent radius of slot ($w_{mi}/4$; w_{mi} is slot width).

c : velocity of light.

$\beta=\omega\sqrt{\varepsilon\mu}$: phase constant.

$I_{nj}(s'_{nj}) \equiv I_J(s'_J)$: current at a point of s'_{nj}.

$m_{nj}(s'_{nj}) \equiv m_J(s'_J)$: magnetic current at a point of s'_{nj}.

$\bar{Q}_{mi\ nj}(s_{mi},s'_{nj}) \equiv \bar{Q}_{IJ}(s_I,s'_J)$: a vector which extends from s_{mi} on the mi-th element axis to s'_{nj} on the nj-th element axis.

(B) Green's function

$$P_{mi\ nj}(s_{mi},s'_{nj}) \equiv P_{IJ}(s_I,s'_J)$$

$$= \frac{e^{-j\beta r_{IJ}(s_I,s'_J)}}{r_{IJ}(s_I,s'_J)}, \qquad \text{(A-8)}$$

where

$$r_{IJ}(s_I,s'_J) = [|\bar{Q}_{IJ}(s_I,s'_J)|^2 + a_I^2]^{\frac{1}{2}}. \qquad \text{(A-9)}$$

$$V_{IJ}(s_I,s'_J) \equiv \frac{1}{4\pi} P_{IJ}(s_I,s'_J). \qquad \text{(A-10)}$$

$$\bar{P}_{IJ}(s_I,s'_J) \equiv P_{IJ}(s_I,s'_J)\ \bar{Q}_{IJ}(s_I,s'_J). \qquad \text{(A-11)}$$

(C) Distance between two points along the m-th wire (slot)

$$D_{mi\ mk}(s_{mi},\eta_{mk}) \equiv D_{IK}(s_I,\eta_K) = d_I + s_I - d_K - \eta_K. \qquad \text{(A-12)}$$

(D) Auxiliary functions

$$\zeta_{mi\ nj}(0,s'_{nj}) \equiv \zeta_{IJ}(0,s'_J)$$

$$= [\hat{s}'_J-(\hat{s}_I\cdot\hat{s}'_J)\hat{s}_I]\cdot\bar{Q}_{IJ}(0,s'_J). \qquad \text{(A-13)}$$

$$\tau_{mi\ nj}(0,s'_{nj}) \equiv \tau_{IJ}(0,s'_J)$$

$$= (\hat{s}_I\times\ \hat{s}'_J)\cdot\bar{Q}_{IJ}(0,s'_J). \qquad \text{(A-14)}$$

$$q_{mi\ nj}(0,s'_{nj}) \equiv q_{IJ}(0,s'_J)$$

$$= r^2_{IJ}(0,s'_J) - [\ \bar{Q}_{IJ}(0,s'_J)\cdot\hat{s}_I]^2. \qquad \text{(A-15)}$$

$$\begin{bmatrix}\eta_{mk\ mi\ nj}(\alpha_{mk},s_{mi},s'_{nj})\\ \xi_{mk\ mi\ nj}(\alpha_{mk},s_{mi},s'_{nj})\end{bmatrix}$$

$$\equiv \begin{bmatrix}\eta_{KIJ}(\alpha_K,s_I,s'_J)\\ \xi_{KIJ}(\alpha_K,s_I,s'_J)\end{bmatrix}$$

$$= \frac{1}{4\pi q_{KJ}(0,s'_J)}\ [\ \bar{P}_{KJ}(\alpha_K,s'_J)\cdot\hat{s}_K\cos\beta D_{IK}(s_I,\alpha_K)$$

$$+ j\sin\beta D_{IK}(s_I,\alpha_K)e^{-j\beta r_{KJ}(\alpha_K,s'_J)}]\ \begin{bmatrix}\zeta_{KJ}(0,s'_J)\\ \tau_{KJ}(0,s'_J)\end{bmatrix}$$

$$\text{(A-16)}$$

$$\begin{bmatrix} \eta^{\#}_{mk\ mi\ nj}(\alpha_{mk}, s_{mi}, s'_{nj}) \\ \\ \xi^{\#}_{mk\ mi\ nj}(\alpha_{mk}, s_{mi}, s'_{nj}) \end{bmatrix}$$

$$\equiv \begin{bmatrix} \eta^{\#}_{KIJ}(\alpha_K, s_I, s'_J) \\ \\ \xi^{\#}_{KIJ}(\alpha_K, s_I, s'_J) \end{bmatrix}$$

$$= \frac{1}{4\pi q_{KJ}(0, s'_J)} [\ \bar{P}_{KJ}(\alpha_K, s'_J) \cdot \hat{s}_K \sin\beta D_{IK}(s_I, \alpha_K)$$

$$- j\cos\beta D_{IK}(s_I, \alpha_K) e^{-j\beta r_{KJ}(\alpha_K, s'_J)}\ ] \begin{bmatrix} \tau_{KJ}(0, s'_J) \\ \\ \zeta_{KJ}(0, s'_J) \end{bmatrix} \tag{A-17}$$

CHAPTER 2
Square Spiral Antenna*

2.1. INTRODUCTION

A square spiral antenna, Fig.2.1, is regarded as a counterpart of a round spiral antenna, which will be discussed in Chapter 3. The radiation mechanism of the square spiral antenna is usually explained by using the current-band theory [1] proposed by J. A. Kaiser. The point of view taken in the current-band theory is as follows: a two-wire spiral antenna behaves as though it were a two-wire transmission line, transforming itself into a radiating structure or antenna. The current-band theory qualitatively gives us an intuitive concept for designing the spiral antenna, but it cannot quantitatively give us the radiation characteristics. Therefore, it is worthwhile to consider the radiation characteristics of the square spiral antenna based on the current distribution.

In this chapter we employ Mei's integral equation, and apply the point-matching method to it for the determination of the current distribution. It will be found that the integration of the kernel is easily carried out by the property of the geometry of the square spiral antenna.

The numerical results of the radiation characteristics for a small square spiral antenna are presented in Section

---

* From papers (p-3),(p-16),(c-1),and (c-7) listed at the end of this monograph.

2.4. The word "small" is given to the spiral whose outer circumference is of the order of two wavelengths. We refer to the effects of the wire radius on the input impedance. In addition, we describe the effects of the arm bend on the radiation field, especially on the axial ratio. An angle coverage of a circularly polarised wave is illustrated for two principal planes. Subsequently, we also investigate the antenna characteristics of a large square spiral antenna for use over a wide frequency range.

2.2. ANTENNA GEOMETRY

The geometry of a two-wire square spiral antenna is shown in Fig.2.1. Arms A and B of the spiral are composed of some linear filaments wound in the X-Y plane. The n-th filament length l_n is defined as

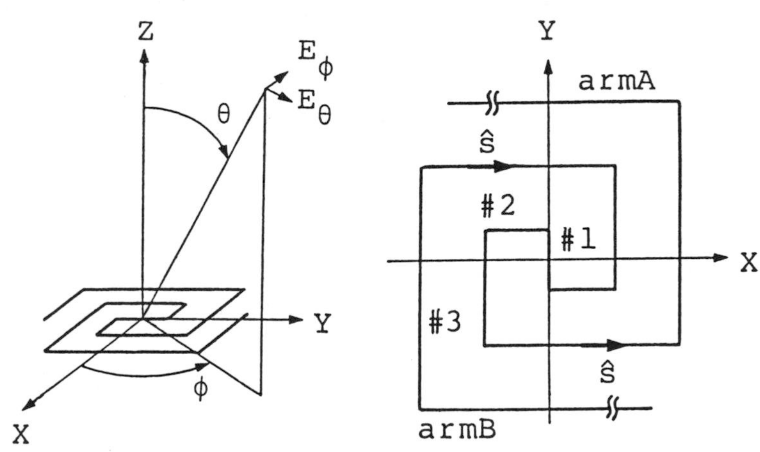

Fig.2.1 Configuration and co-ordinate system.

$$l_n = \begin{cases} a & \text{for } n=1 \\ 2a(n-1) & \text{for } n=2,3,\cdots \end{cases} \quad . \tag{2.1}$$

The position vector from the origin to an arbitrary point on the n-th filament is given by

$$\bar{r}(s) = [-1]^q \Big[[a(n-1)\sin\frac{n\pi}{2} + \{s-a(n-1)^2\} \cos\frac{n\pi}{2}]\ \hat{x}$$

$$+ [-a(n-1)\cos\frac{n\pi}{2} + \{s-a(n-1)^2\} \sin\frac{n\pi}{2}]\ \hat{y} \Big]$$

$$\begin{cases} q=0 & \text{on the arm A} \\ q=1 & \text{on the arm B ,} \end{cases} \tag{2.2}$$

where $\hat{x}$ and $\hat{y}$ are unit vectors in the rectangular co-ordinates, and s is the distance measured along the antenna arm from the feed point. Trigonometric functions are used as indicators showing 0 and ±1.

The tangential unit vector along the n-th filament is expressed by

$$\hat{s} = \cos\frac{n\pi}{2}\ \hat{x} + \sin\frac{n\pi}{2}\ \hat{y} \tag{2.3}$$

on the arms A and B. The antenna is fed from the origin by a delta function generator of voltage V_0.

2.3. NUMERICAL METHOD

The current distribution I(s') along the spiral arm may be determined by using the techniques mentioned in Chapter 1. In this section we employ Mei's integral equation (1.26) and study how to apply it to the square spiral antenna.

The integration of the left side of eqn.(1.26) may be reduced to an integration over the half-arm length because of the symmetry of the spiral geometry with respect to the origin. Taking account of the arm bends and the orthogonal property among the arm filaments, we transform eqns.(1.27), (1.28), and (1.29) into

$$\pi_1(s,s') = G(s,s')\hat{s}\cdot\hat{s}' , \qquad (2.4)$$

$$\pi_2(s,s') = \sum_{i=1}^{Q} G(c_i,s')(\hat{c}_i^+ - \hat{c}_i^-)\cdot\hat{s}' \cos\beta(s-c_i) , \qquad (2.5)$$

$$\pi_3(s,s') = \int_0^s g(\xi,s') \cos\beta(s-\xi)d\xi , \qquad (2.6)$$

where

$$g(\xi,s') = \begin{cases} 0 & \text{for} \quad m_\xi + n_{s'} = \text{even number} \\ \dfrac{\partial G(\xi,s')}{\partial s'} = -G(\xi,s')\dfrac{1+j\beta D(\xi,s')}{D^2(\xi,s')} \cdot \dfrac{(x'-\xi_x)+\dfrac{dy'}{dx'}(y'-\xi_y)}{\sqrt{1+\left(\dfrac{dy'}{dx'}\right)^2}} & \text{for} \quad m_\xi + n_{s'} = \text{odd number} , \end{cases} \qquad (2.7)$$

$$G(s,s') = \frac{\exp[-j\beta D(s,s')]}{4\pi D(s,s')} \,, \tag{2.8}$$

$$D(s,s')=|\bar{r}(s)-\bar{r}(s')| \,, \tag{2.9}$$

$s'(x',y')$ and $\xi(\xi_x,\xi_y)$ are the distances measured along the antenna arm from the feed point, c_i is the value of ξ at the i-th bend, and $\hat{c}_i^-$ and $\hat{c}_i^+$ are unit vectors before and after the i-th bend, respectively. The summation limit Q in eqn.(2.5) is the number of bends the current has passed before it reaches s. m_ξ and $n_{s'}$ in eqn.(2.7) are the integers labelled to the filaments on which the points whose distances are ξ and s' exist, respectively.

The integration of $\pi_1(s,s')$ becomes singular as s' approaches s. However, the singularity is eliminated by using a well-known treatment [2], in which the surface current is replaced by a line current along the centre of the antenna arm. In addition, it should be noted that singularities of $\pi_2(s,s')$ and $\pi_3(s,s')$ which appear as s' approaches c_i in eqn.(2.5) and ξ approaches s' near the bend in eqn.(2.6), respectively, can be eliminated by the following properties [3]:

$$\lim_{\varepsilon\to 0} \int_{c_i-\varepsilon}^{c_i+\varepsilon} I(s')G(c_i,s')[\hat{c}_i^+ - \hat{c}_i^-]\cdot\hat{s}'\cos\beta(s-c_i)ds'=0 \,, \tag{2.10}$$

$$\lim_{\varepsilon\to 0}\left[\int_{c_i-\varepsilon}^{c_i} I(s')\int_{c_i}^{c_i+\varepsilon} g(\xi,s')\cos\beta(s-\xi)d\xi\, ds'\right.$$

$$\left.+\int_{c_i}^{c_i+\varepsilon} I(s')\int_{c_i-\varepsilon}^{c_i} g(\xi,s')\cos\beta(s-\xi)d\xi\, ds'\right]=0\ . \tag{2.11}$$

These properties make the computation simple and indicate that the geometry of the rectangular bend is not an essential factor in the operation of a thin wire antenna.

So far, we have discussed the kernel of Mei's integral equation. Now, we apply the point-matching method to eqn.(1.26), using a pulse function as the expansion function. The current is expressed as $I(s')=I_{\Delta s'}$ for $s'\in$ a small segment $\Delta s'$ and $I(s')=0$ for $s'\notin\Delta s'$. This choice leads the integration of $g(\xi,s')$ to a single integration:

$$\int_{\Delta s'} I(s')\left[\int_0^s g(\xi,s')\cos\beta(s-\xi)d\xi\right]ds'$$

$$=I_{\Delta s'}\int_0^s [G(\xi,s')]_{\Delta s'}\cos\beta(s-\xi)d\xi\ . \tag{2.12}$$

Thus, the calculation of the left side of eqn.(1.26) is completed together with the single integrations of eqns.(2.4) and (2.5) over the antenna length.

Once the current distribution is determined, two components of the electric field at a far-field point (R,θ,ϕ) are given by

$$E_\theta = C \cos\theta \sum_{n=1}^{N} \int_{\substack{n\text{-th}\\ \text{filament}}} I(s') \left[\cos\frac{n\pi}{2}\cos\phi + \sin\frac{n\pi}{2}\sin\phi \right] \cdot \cos\beta\gamma \, ds' , \tag{2.13}$$

$$E_\phi = C \sum_{n=1}^{N} \int_{\substack{n\text{-th}\\ \text{filament}}} I(s') \left[\sin\frac{n\pi}{2}\cos\phi - \cos\frac{n\pi}{2}\sin\phi \right] \cdot \cos\beta\gamma \, ds' , \tag{2.14}$$

where

$$C = -j \frac{\omega\mu}{2\pi R} e^{-j\beta R} , \tag{2.15}$$

$$\gamma = \Big[\big[a(n-1)\sin\frac{n\pi}{2} + \{s'-(n-1)^2 a\}\cos\frac{n\pi}{2} \big] \cos\phi + \big[-a(n-1)\cos\frac{n\pi}{2} + \{s'-(n-1)^2 a\}\sin\frac{n\pi}{2} \big] \sin\phi \Big] \sin\theta , \tag{2.16}$$

and the summation limit N is the total number of the arm filaments.

2.4. SMALL SQUARE SPIRAL ANTENNA

We analyse a small square spiral antenna in which the

number of filaments is 9 and the first filament length l_1=a=0.3077 cm, with the frequency fixed at 3 GHz (wavelength λ=10 cm). The outer circumference of the spiral is about 2λ, and the spiral radiates in the first mode [1]; i.e., the dominant radiation is produced from the current along a mean circumference of one wavelength.

Fig.2.2 shows the current distribution along the spiral arm under the condition that the driving voltage at the input is 1 volt. The convergence of the current distribution reaches a satisfactory state with 33 subdivisions (segments) per arm. It is found that the current travels with a phase velocity which is nearly equal to the velocity of light.

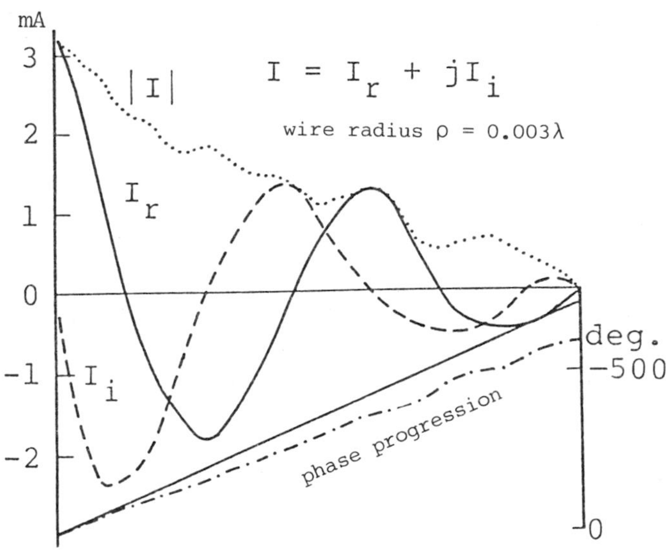

Fig.2.2 Current distribution.

Fig.2.3 shows the input impedance as a function of the wire radius ρ ranging from 0.001λ to 0.01λ [4]. The resistance value tends to decrease as the wire radius is increased, while the reactance value remains about zero. At a wire radius of 0.01λ the input impedance is nearly a pure resistive value of 180 ohms. It is interesting to note that the input impedance is close to 60π ohms which is that of a self-complementary antenna, although the square spiral is not exactly self-complementary. (The ratio of an equivalent strip width 4ρ to the spacing width between the neighbouring filaments is 1 at $\rho=0.0077\lambda$.)

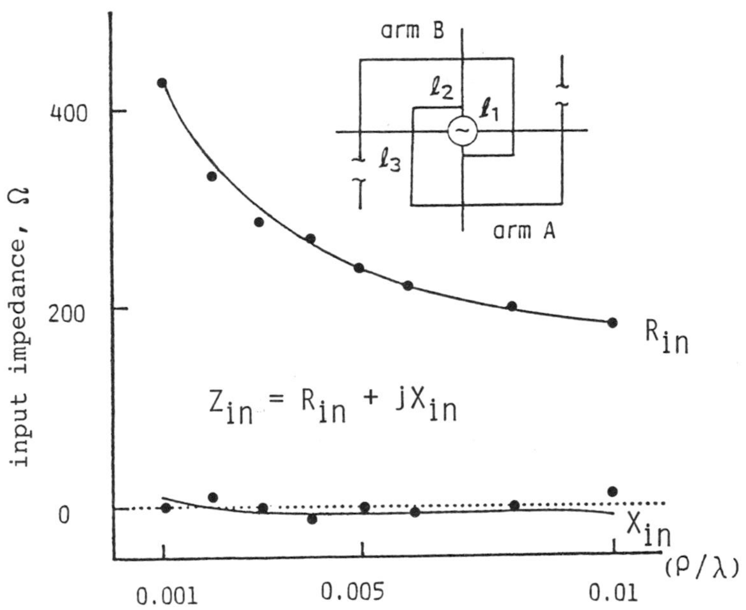

Fig.2.3 Input impedance as a function of wire radius.

theoretical ————— experimental • • • • •

For conformal applications such as on vehicles, it is sometimes necessary to bend the spiral arms over the body of the object and it is interesting, therefore, to consider the effects of the bend on the radiation characteristics [4] of the spiral. Fig.2.4 shows an antenna which is bent symmetrically with respect to the Y-axis, where the bend angle is designated as Ψ. In the present analysis the wire radius ρ is taken to be 0.005λ.

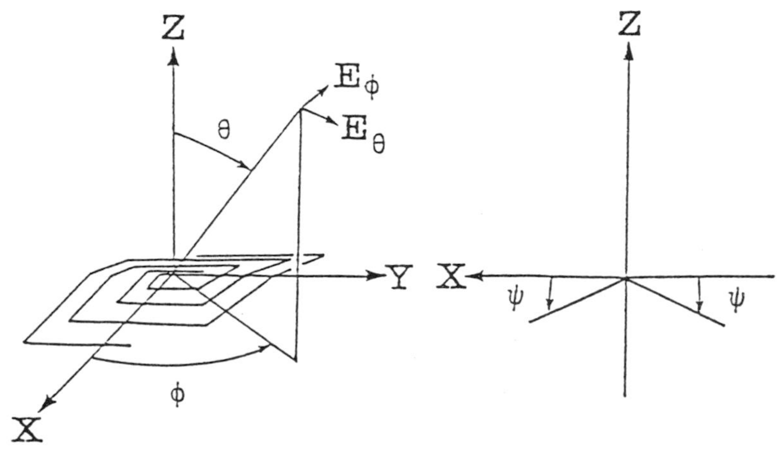

Fig.2.4 Configuration of bent square spiral antenna.

Fig.2.5 shows the amplitudes of the current distributions as a function of the bend angle. The smooth decay of the current ensures a radiation wave of circular polarisation, and the similarity among the current distributions leads to the fact that the input impedance remains relatively constant. The input impedance shows nearly a pure resistance, and the value for $\Psi=0°$ is 250 ohms, while that for $\Psi=30°$ is 290 ohms.

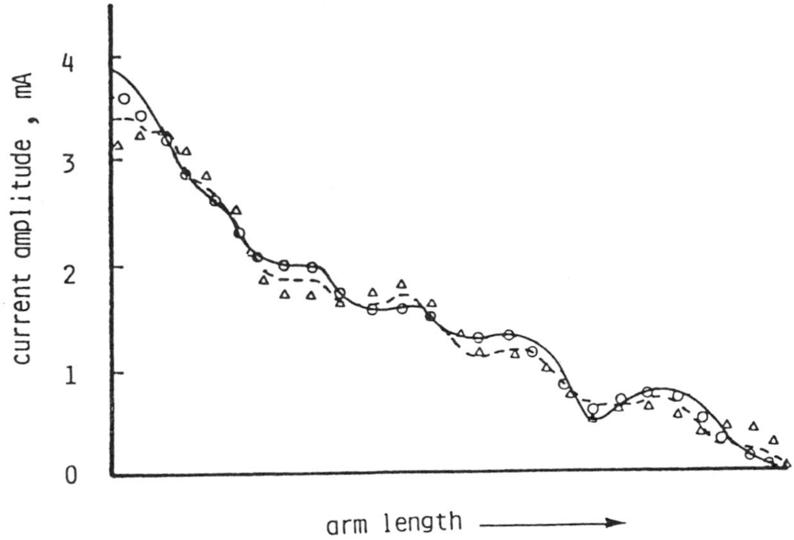

Fig.2.5 Current amplitudes as a function of bend angle.

——— $\Psi = 0°$ ○ ○ ○ ○ ○ $\Psi = 20°$

- - - - - $\Psi = 30°$ △ △ △ △ △ $\Psi = 40°$

For use as a unidirectional antenna, the spiral is usually supported by a cavity with the absorber at the - Z-side. The radiation toward - Z-side is absorbed without affecting the radiation toward + Z-side. Therefore, it is important to evaluate the radiation field at the + Z-side. Fig.2.6 shows the axial ratio as a function of θ at the + Z-side. Fortunately, in the $\phi=90°$

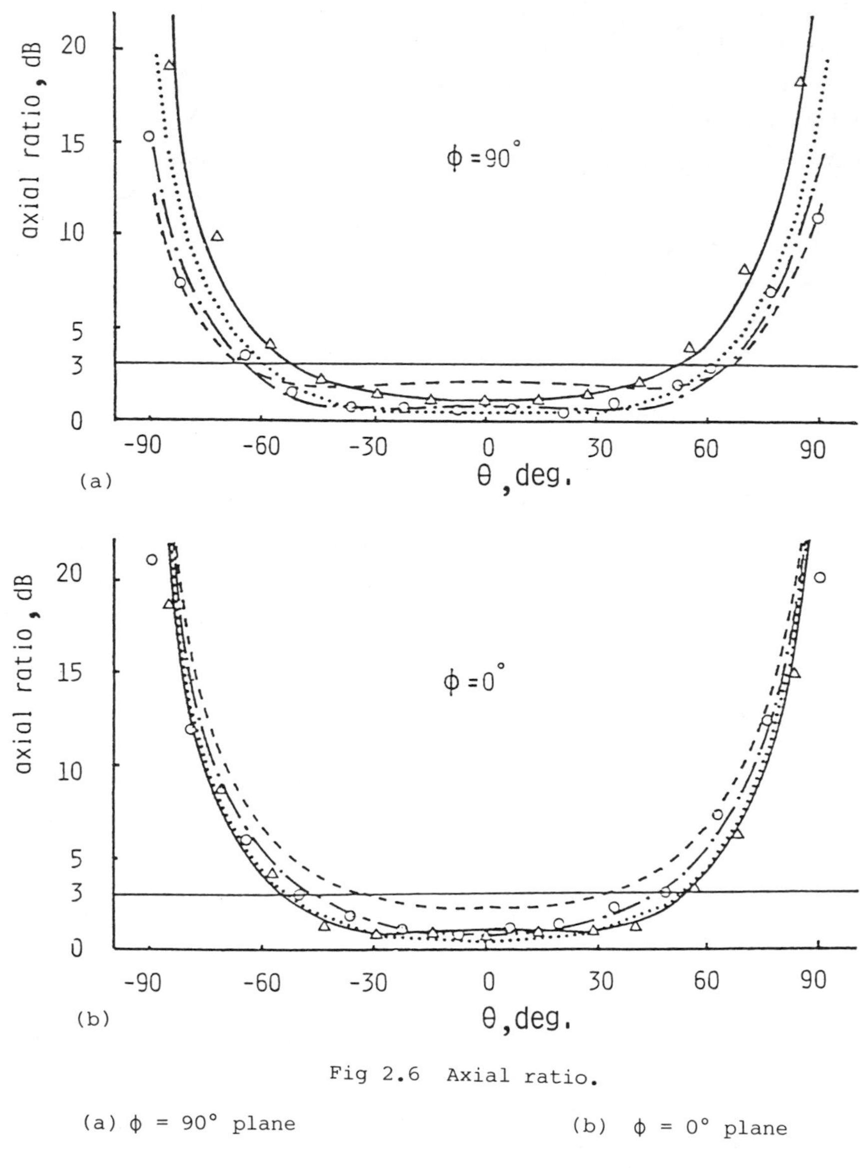

Fig 2.6 Axial ratio.

(a) φ = 90° plane (b) φ = 0° plane

theoretical:

——— Ψ = 0° ········· Ψ =10°

—·— Ψ =20° - - - - - Ψ =30°

experimental:

△ △ △ △ △ Ψ = 0°

○ ○ ○ ○ ○ Ψ =20°

plane, the angle coverage of the circularly polarised wave tends to increase with an increase in the bend angle. The angle range with an axial ratio of less than 3 dB widens from 104° to 130°, as the bend angle Ψ is increased from 0° to 20°. In the ϕ=0° plane, a significant change in the angle coverage of the circularly polarised wave is not observed up to Ψ=20°.

The E_θ component for the bent spiral is always found to exist at θ=±90°, unlike that for a planar spiral, as shown in Fig.2.7. This is a factor in the wide-angle coverage of the circularly polarised wave in the + Z-axis direction.

Further calculation shows that even when the bend angle is changed from 0° to 30°, the power gain remains nearly constant. The power gain is about 3.9 dB in the + Z-axis direction.

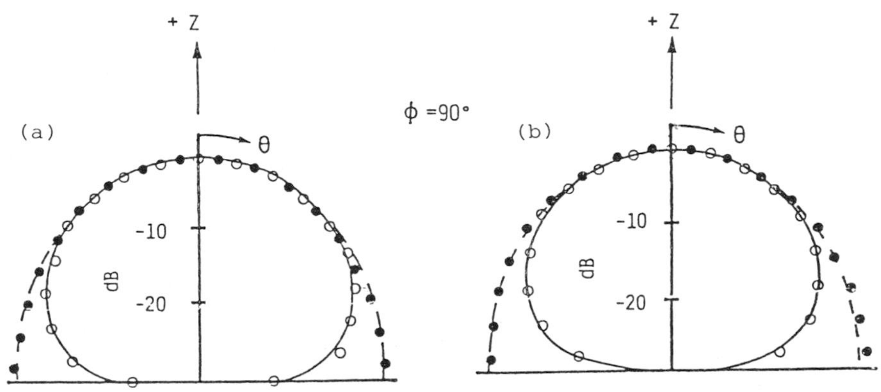

Fig.2.7 Radiation patterns.

(a) Ψ =20°

(b) Ψ = 0°

theoretical

———————— E_θ

- - - - - - - - - - E_ϕ

experimental

○ ○ ○ ○ ○ E_θ

● ● ● ● ● E_ϕ

2.5. LARGE SQUARE SPIRAL ANTENNA

The previous antenna described in Section 2.4 is relatively narrowband. We next consider a square spiral antenna for use over a wide frequency range [5]. The dimensions of an antenna to be considered here are as follows: the half-arm length L=25.7 cm (the number of turns$\simeq$4), the first filament length a=1 mm, the wire radius ρ=0.2 mm. We calculate the antenna characteristics over a frequency range of 3 to 7 GHz.

Fig.2.8 shows the numerical results of the current distributions and the corresponding radiation patterns. The situation shown in Fig.2.8(a) is regarded as a transient case where the spiral is about to establish a radiation beam of circular polarisation. The coincidence between the E_θ and E_ϕ levels in the radiation pattern is not yet observed. Fig.2.8(b) is a typical current distribution and its phase progression when the antenna radiates a circularly polarised wave. It is found that the phase progression is nearly equal to the free-space phase progression.

A comparison of the amplitudes of current distributions is shown in Fig.2.8(c), where the abscissa is the arm length normalised to the operating wavelength. The decay of each current from the feed point to the location of the first mode radiation in terms of the current band theory is found to be similar at frequencies of more than 4 GHz. This characteristic makes the radiation pattern uniform over a wide frequency range, as shown in Fig.2.8(d). We observe that the E_ϕ-component has a wider pattern than the E_θ-component (the half-power beamwidth of the E_θ-component is about $\pm 40°$).

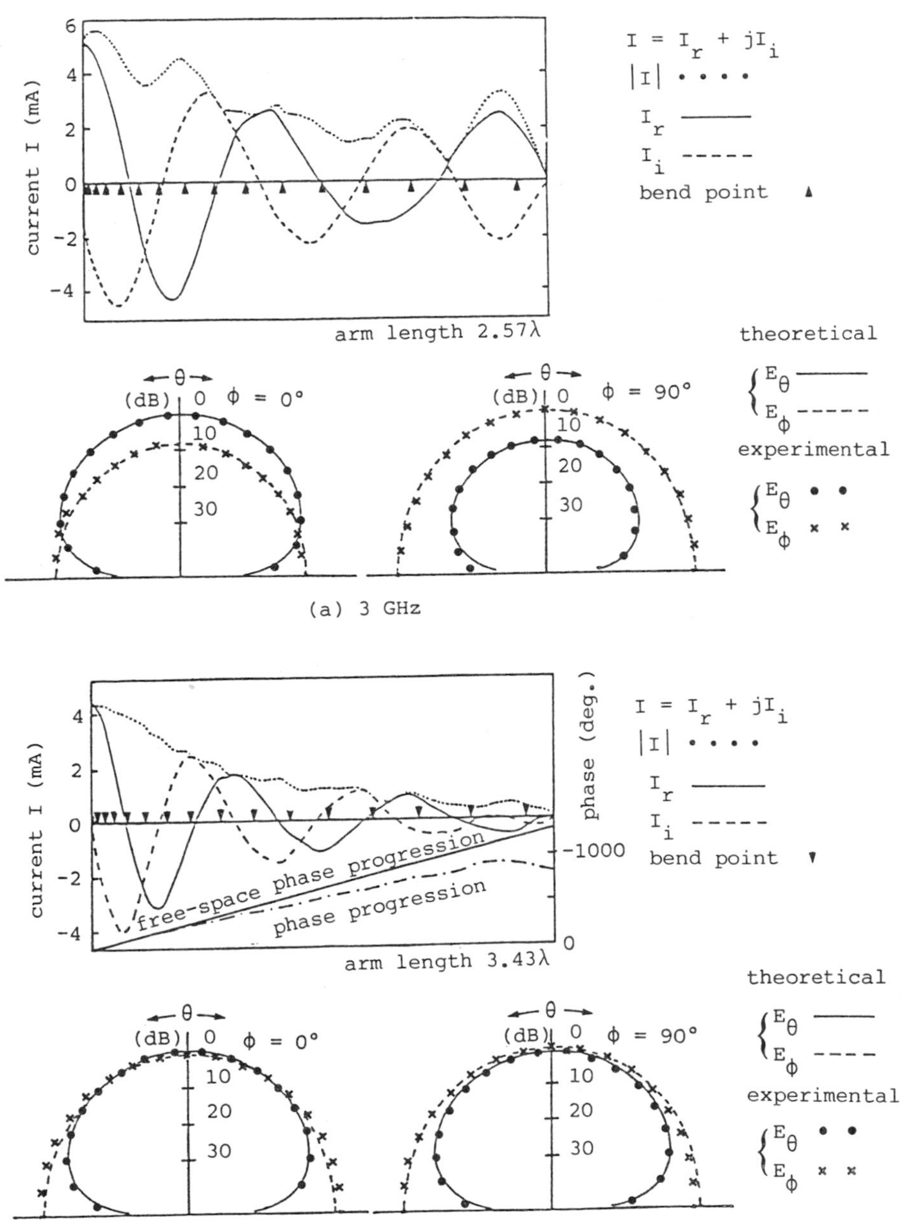

Fig.2.8 (a),(b) Frequency characteristics of current distribution and radiation pattern in Z-axis cut.

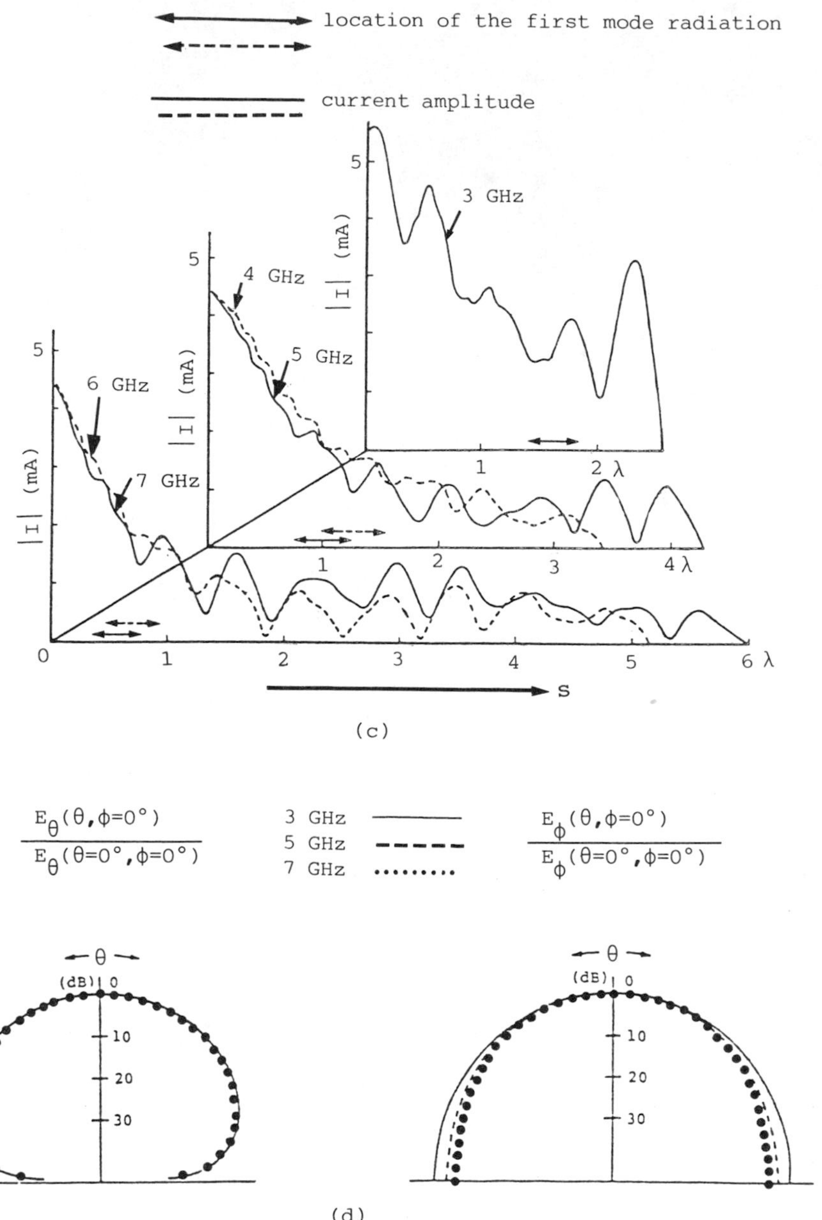

Fig.2.8 (c),(d) Frequency characteristics of current distribution and radiation pattern in Z-axis cut.

Fig.2.9 shows the theoretical and experimental results of the power gain and the axial ratio. The power gain is in a range of 3.6 dB to 4.9 dB . Although the axial ratio is 8.6 dB at 3 GHz, its rapid improvement is observed over a frequency range of 3 to 4 GHz.

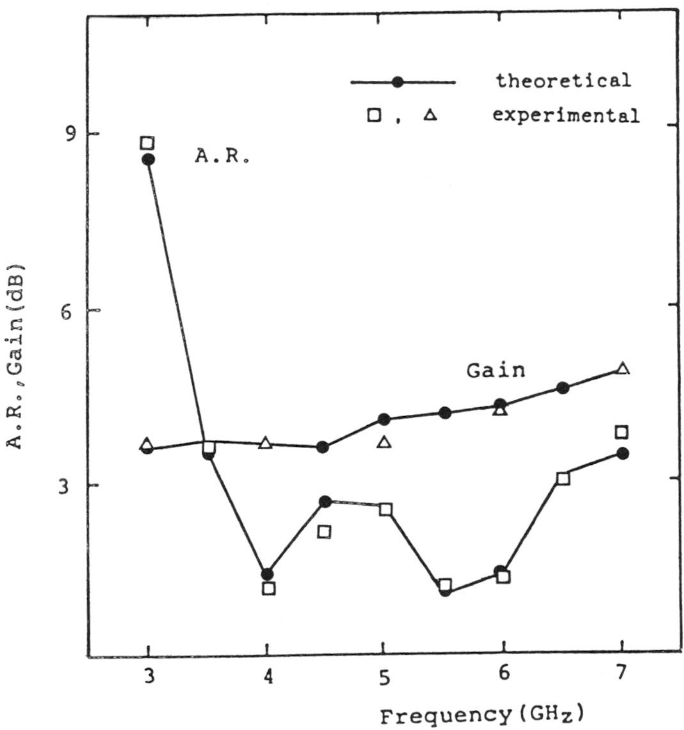

Fig.2.9 Power gain and axial ratio versus frequency.

Fig.2.10 shows the input impedance when the frequency is changed. The input impedance becomes stable at frequencies of more than 3.5 GHz with nearly a pure resistance of 220 ohms. Although this resistance value seems to be high for matching with the characteristic impedance of a coaxial line, we can easily settle the impedance matching problem by using a printed balun circuit [6]. A desirable performance is that the printed balun circuit operate over a wideband frequency.

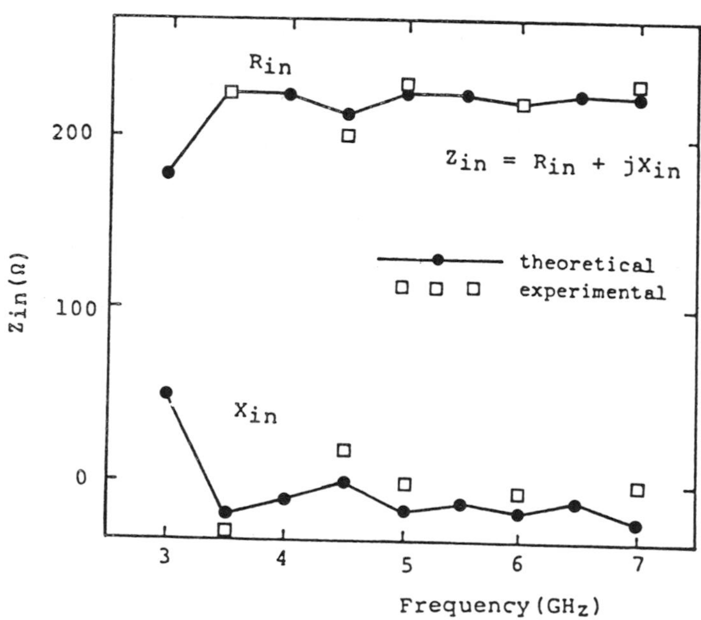

Fig.2.10 Input impedance versus frequency.

2.6. CONCLUSION

In Section 2.4 it has been revealed that the input impedance of a small square spiral antenna decreases as the wire radius is increased. A value close to 60π ohms is obtained when the ratio of the equivalent strip width to the spacing width between the neighbouring filaments is about unity, although the square spiral is not exactly self-complementary.

The radiation characteristics when the arm of the small square spiral is bent have also been investigated. The spiral has the bend angles in which the angle coverage of a circularly polarised wave is widened in one principal plane, while maintaining the inherent input impedance and power gain.

In Section 2.5 we have evaluated the antenna characteristics of a large square spiral antenna. The evaluation covers the current distribution over a frequency range of 3 to 7 GHz. It also covers the power gain, axial ratio and input impedance over the same frequency range. The following conclusions are reached; the decay in the main region of the current distribution maintains a similar form in spite of the change of frequency, resulting in the wideband characteristics of the radiation.

REFERENCES TO CHAPTER 2

[1] Kaiser, J.A., "The archimedean two-wire spiral antenna", IRE Trans., AP-8, 1960, pp.312-323.

[2] Yeh, Y.S. and Mei, K.K., "Theory of conical equiangular spiral antennas. part 1-numerical technique", IEEE Trans.,AP-15, 1967, pp.634-639.

[3] Lee, S.H. and Mei, K.K., "Analysis of zigzag antennas", IEEE Trans., AP-18, 1970, pp.760-764.

[4] Nakano, H., Yamauchi, J and Nogami, K, "Effects of wire radius and arm bend on a rectangular spiral antenna", IEE, Electron Lett., vol.19, No.23, 1983, pp.957-958.

[5] Nakano, H., and Yamauchi, J. and Koizumi, H., "The theoretical and experimental investigations of the two-wire square spiral antenna", Trans.IECE.Japan, E63, No.5, 1980, pp.337-343.

[6] Bawer, R., and Wolfe, J.J, "A printed circuit balun for use with spiral antennas", IRE Trans. MTT-8, 1960, pp.319-325.

CHAPTER 3
Two-Wire Round Spiral Antenna*

3.1. INTRODUCTION

One of the factors that determine the operation bandwidth of a spiral antenna is the axial ratio of the radiation field. The lower limit of the bandwidth is dependent on the outer circumference of the spiral, while the upper limit is dependent on the configuration near the feed point.

It is pointed out that, as the frequency becomes lower, the reflected currents from the arm end increase and the axial ratio deteriorates. Therefore, suppressing or reducing the reflected currents is fundamental for obtaining a circularly polarised wave. A countermeasure such as coating the arms with lossy material is often used to suppress the reflected currents, but this chemical countermeasure is troublesome and is not necessarily economical.

This chapter is devoted to the improvement of the axial ratio of the radiation field from a spiral antenna without chemical countermeasures. We introduce small zigzag elements on the outermost arms of the spiral in order to reduce the reflected currents. On the basis of the radiation field we evaluate the axial ratio by using eqn.(1.72), and determine a spiral antenna with the optimised zigzag elements. The radiation characteristics

* From papers (p-8),(p-11),(c-2),(c-4),(c-8) and (c-9) listed at the end of this monograph.

are compared with those of a smooth round spiral antenna without zigzag elements.

3.2. ANTENNA GEOMETRY AND NUMERICAL METHOD

Fig.3.1 shows the configuration of a two-wire round spiral antenna. The two arms are wound symmetrically with respect to the origin, each arm consisting of smooth and zigzag sections. The radius of the smooth section is defined by Archimedean spiral function $r=a\phi_s$ (a=spiral constant, ϕ_s=winding angle) starting from an angle ϕ_{st} with the first segment length e_{st}. The zigzag section is made only in the outermost arm and exists on an imaginary expansion of the Archimedean spiral curve. The zigzag is characterised by an apex angle τ and an element length e [1].

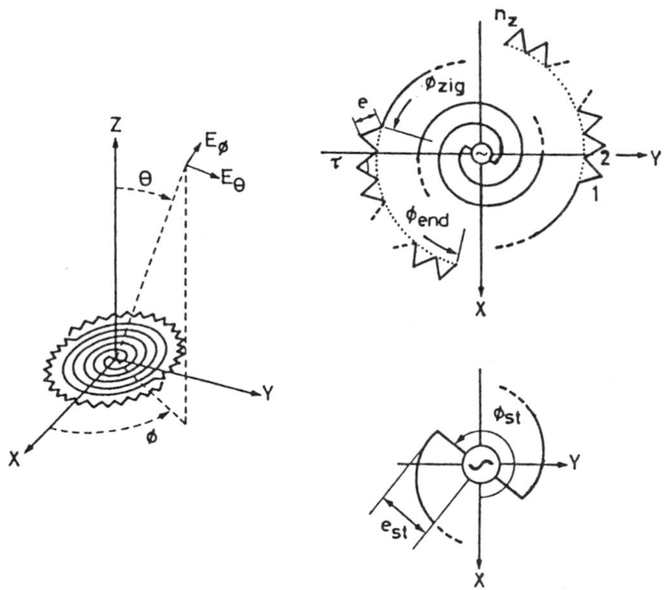

Fig.3.1 Configuration of spiral antenna.

We determine the current distribution of the spiral antenna by using Nakano's integral equation (1.63), and calculate the electric field at a far-field point (R,θ,ϕ) on the basis of the determined current distribution $I_j(s_j')$. Two components of the electric field are

$$E_\theta = -j \frac{\omega\mu}{2\pi R} e^{-j\beta R} \cdot \sum_j \int_{j\text{-th}} I_j(s_j')\cos[\beta\hat{R}\cdot(\bar{r}_j + s_j'\hat{s}_j')]\hat{s}_j'\cdot\hat{\theta}ds_j' , \qquad (3.1)$$

$$E_\phi = -j \frac{\omega\mu}{2\pi R} e^{-j\beta R} \cdot \sum_j \int_{j\text{-th}} I_j(s_j')\cos[\beta\hat{R}\cdot(\bar{r}_j + s_j'\hat{s}_j')]\hat{s}_j'\cdot\hat{\phi}ds_j' , \qquad (3.2)$$

where $\hat{R}$, $\hat{\theta}$ and $\hat{\phi}$ are the unit vectors in the spherical co-ordinates, and $\bar{r}_j$ is a position vector from the origin to a starting point of the j-th element, as shown in Fig.3.2.

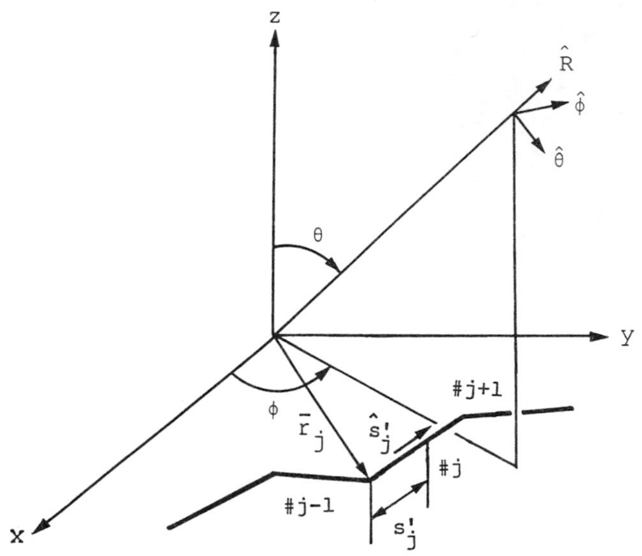

Fig.3.2 Co-ordinates for calculation of radiation field.

3.3. ANTENNA PARAMETERS [2]

The spiral antenna needs an outermost circumference of more than one wavelength in order to radiate a circularly polarised wave. For the operation of the spiral antenna at frequencies of more than 3 GHz, we choose the parameters of the spiral as follows: the first segment angle ϕ_{st}=4.17 rad; the first segment length e_{st}=3 mm; the spiral constant a=0.72 mm/rad; the starting angle of the zigzag section ϕ_{zig}=29.42 rad; the element length of the zigzag section e=3 mm; the number of the zigzag elements n_z=32; and the wire radius ρ=0.3 mm.

The apex angle τ should be determined so that the axial ratio may be improved effectively. Fig.3.3 shows the behaviour of the axial ratio as a function of the apex angle at 3 GHz. When the apex angle is 180°, the antenna corresponds to a smooth round spiral antenna. For the range of this numerical result, the best axial ratio is obtained when the zigzag section is made outside the imaginary expansion of the Archimedean spiral curve with an apex angle of 60°. In this case, the winding angle of the arm end is ϕ_{end}=31.6 rad. It is observed that an axial ratio of 2.0 dB for an apex angle of 180° is improved to 0.7 dB for an apex angle of 60°.

On the basis of the above-mentioned results, we further investigate a spiral antenna having 32 zigzag elements of an apex angle of 60°, and compare the radiation characteristics with those of the corresponding smooth round spiral antenna with the same angle of the arm end (ϕ_{end}=31.6 rad). The testing frequency range is taken between 2.6 GHz and 6 GHz in the following section.

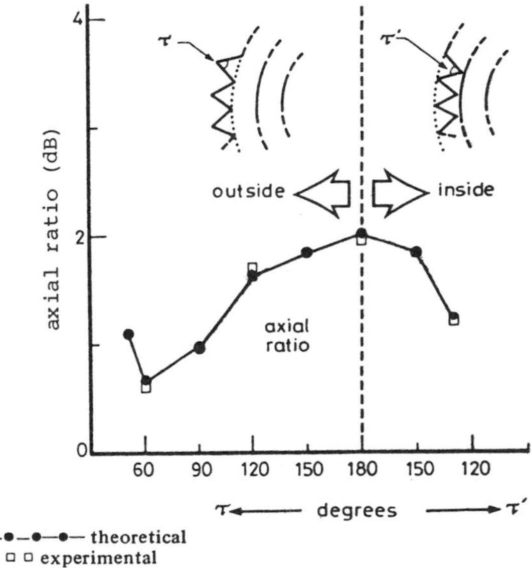

Fig.3.3 Axial ratio as a function of apex angle τ at 3 GHz.

3.4. CURRENT DISTRIBUTION AND RADIATION PATTERN IN THE Z-AXIS CUT [1]

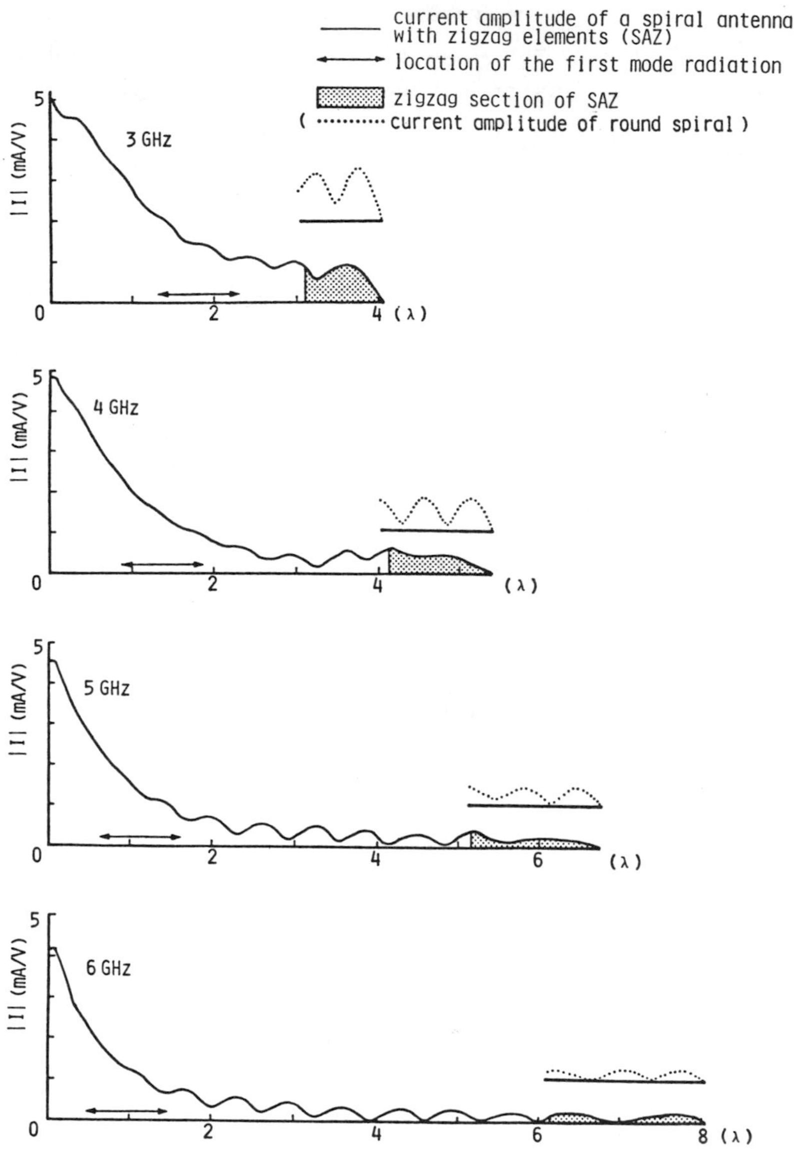

Fig.3.4 Amplitude of current distribution.

We subdivide the antenna arm into 135 elements in order to obtain the current distribution. Fig.3.4 shows the amplitudes of the current distributions over a frequency range of 3 to 6 GHz. For comparison, the current distribution near the arm end of the smooth round spiral is also shown. The abscissa in the Figure is the arm length normalised to the operating wavelength. It is found that at lower frequencies the current standing wave near the arm end is considerably reduced, as compared with that of the smooth round spiral. This is closely related to the improvement of the axial ratio as will be mentioned in Section 3.6.

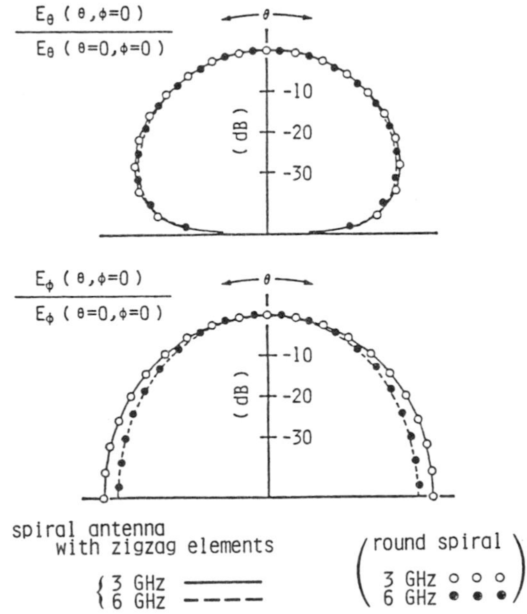

Fig.3.5 Theoretical radiation pattern in the Z-axis cut as a function of frequency.

Fig.3.5 shows the theoretical radiation patterns of the two spiral antennas as a function of frequency. The E_ϕ component is somewhat dependent on the frequency, while the E_θ component shows a stationary pattern with the half-power beamwidth of about ± 40°, as in the case of a square spiral antenna analysed in Chapter 2. We do not find any significant difference in the E_θ component between the two spiral antennas. This is also true for the E_ϕ component. It should be noted that, as described in Chapter 2, a wideband characteristic of the radiation pattern results from the fact that similar decay of the current over the main region from the feed point to the location of the first mode radiation is maintained regardless of the change in the frequency.

3.5. RADIATION PATTERN IN THE SPIRAL PLANE AND INPUT IMPEDANCE [1]

The radiation field in the spiral plane has only the E_ϕ component; i.e., the E_θ component does not exist in the XY-plane. Fig.3.6 shows a typical radiation pattern of the E_ϕ in the spiral plane and its phase change as a function of the azimuth.

According to a qualitative explanation by a loop model of the current-band theory [3], one degree of the mechanical rotation of the spiral antenna precisely produces a phase change of one degree in the radiation field. The practical antenna, however, somewhat loses this relation in the phase change, though the mechanical rotation of 360° precisely causes the phase change of 360° in the radiation field, as shown in Fig.3.6.

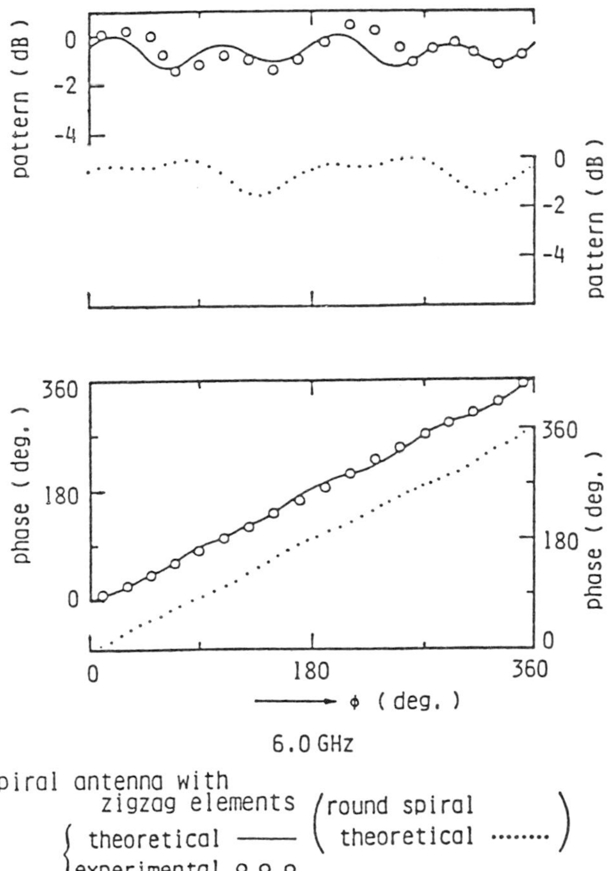

Fig.3.6 Radiation pattern and phase change in the spiral plane.

Table 1 shows a comparison of the E_ϕ pattern variations between the two spiral antennas. The spiral with the zigzag elements shows smaller variation than the smooth round spiral. It is noted that the reduction of the reflected current from the arm end contributes to the omnidirectional radiation in the spiral plane.

The input impedance of each spiral antenna is almost independent of the frequency, as seen from Fig.3.7. The resistance value is of the order of 200 ohms and its variation is only about 30 ohms over a frequency range of 2.6 to 6 GHz, while the reactance value is of the order of 50 ohms with almost the same variation of 30 ohms.

TABLE 1 Pattern variation in the spiral plane, 20 log $[E_{\phi min}/E_{\phi max}]$ (dB).

frequency (GHz)	spiral antenna with zigzag elements ϕ_{end}=31.6 (rad.) τ=60° n_z=32	round spiral antenna ϕ_{end}=31.6 (rad.)
2.6	-2.70	-5.33
3.0	-1.06	-1.87
4.0	-0.79	-2.20
5.0	-0.88	-1.88
6.0	-1.35	-1.45

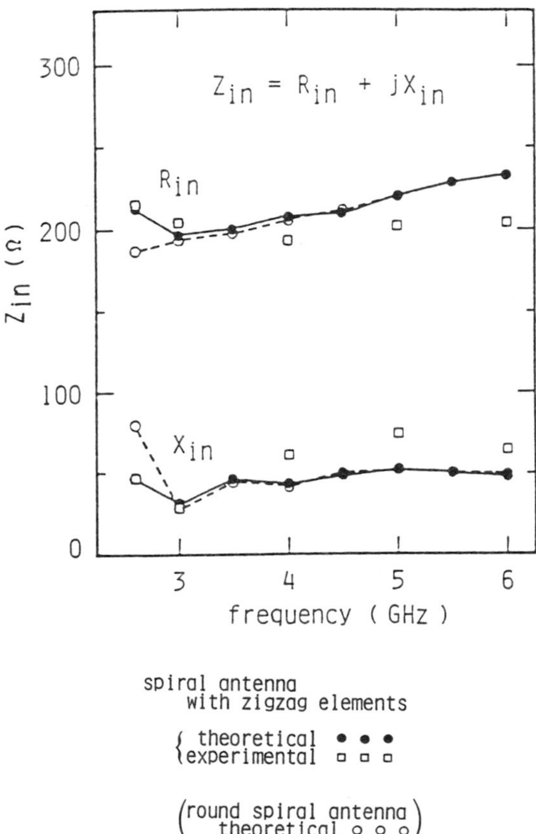

Fig.3.7 Input impedance as a function of frequency.

3.6. POWER GAIN AND AXIAL RATIO [1]

The similarity in the power gain between the two spiral antennas is found in Fig.3.8. The slight increase in the power gain at higher frequencies is attributed to the fact that the half-power beamwidth of the E_ϕ component becomes somewhat narrow as the frequency is increased, as shown in Fig.3.5. The variation of the power gain is only 1.3 dB over a frequency range of 2.6 to 6 GHz.

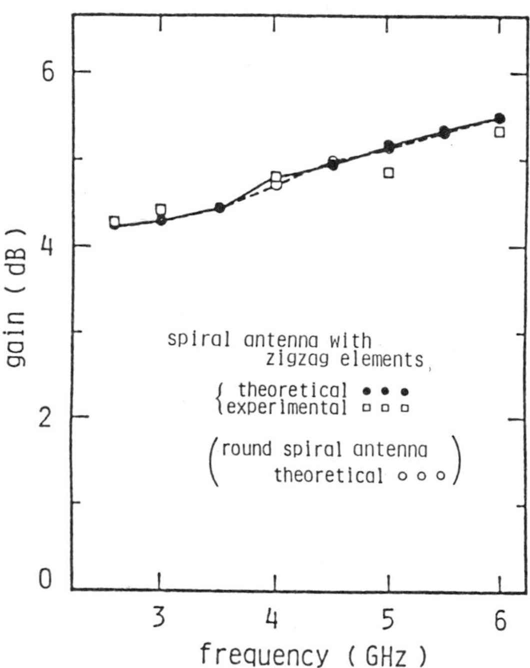

Fig.3.8 Power gain as a function of frequency.

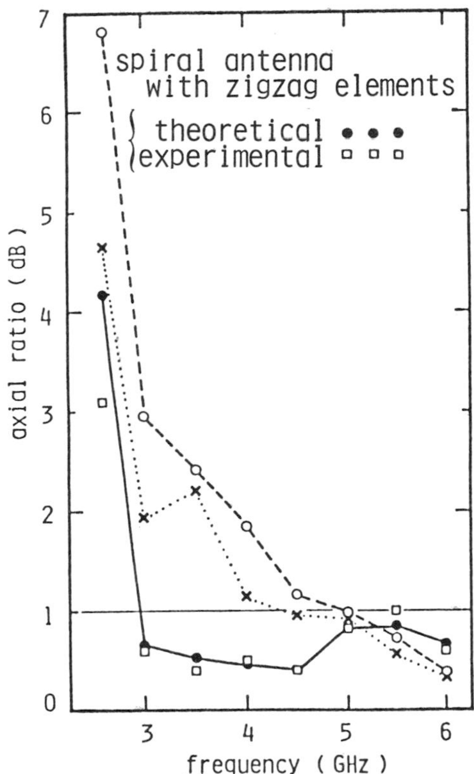

spiral antenna with zigzag elements

ϕ_{end} = 31.6 (rad.) arm length =40.5 (cm)

round spiral antenna

o o o ; ϕ_{end} = 31.6 (rad.) arm length = 35.7 (cm)

× × × ; ϕ_{end} = 33.6 (rad.) arm length = 40.5 (cm)

Fig.3.9 Axial ratio as a function of frequency.

Fig.3.9 shows the axial ratios of the two spiral antennas as a function of frequency. Generally speaking, after the establishment of the first mode radiation, the axial ratio of the smooth round spiral is improved with the increase in the frequency. However, the improvement is not rapid, as shown by a broken line. In contrast, the spiral with the zigzag elements shows a noticeable improvement toward circular polarisation at lower frequencies, by virture of the smooth current distribution resulting from the reduction of the reflected current.

At higher frequencies the difference in the axial ratio between the two spiral antennas is small. In order to understand the reason for this, we should recall that, as the frequency is increased, the remaining current along the outer arm of each spiral antenna becomes considerably small as compared with the current over the main region from the feed point to the location of the first mode radiation. In other words, the axial ratios of the two spiral antennas are characterised by the similar currents, and hence they are close to each other.

For additional information, the axial ratio of another smooth round spiral antenna whose arm length is the same as that of the spiral with the zigzag elements is also shown in Fig.3.9. At lower frequencies we again find the superiority of the spiral with the zigzag elements over the smooth round spiral.

3.7. CONCLUSION

A spiral antenna with zigzag elements exhibits a wideband characteristic of the axial ratio, which is not obtainable in a smooth round spiral antenna, without deteriorating the wideband characteristics of the radiation pattern, input impedance and power gain.

Some comments are made for further information: (i) Experimental data in References [4] and [5] are available to realise the wideband characteristics of the radiation from a spiral antenna backed by a cavity. (ii) Reference [6] is also informative, which describes the frequency bandwidth when a spiral antenna is backed by a conducting plane reflector. (iii) If a spiral element is mounted on a grounded dielectric substrate (a microstrip spiral antenna), the numerical analysis can be made using a technique described in Reference [7].

REFERENCES TO CHAPTER 3

[1] Nakano, H., Yamauchi, J. and Hashimoto, S., "sunflower spiral antenna", Trans. IECE Japan, E64, 12, 1981, pp.763-769.

[2] Nakano, H. and Yamauchi, J., "Characteristics of modified spiral and helical antennas", Proc. IEE, Vol.129, 1982, pp.232-237.

[3] Kaiser, J.A., "The archimedean two-wire spiral antenna", IRE Trans., AP-8, 3, 1960, pp.312-323.

[4] Morgan, T. E., "Spiral antennas for ESM", IEE Proc. Vol.132, Pt.F, No.4, 1985, pp.245-251.

[5] Morgan, T. E., "Spiral antennas for ESM", Proc. IEEE International symposium on antennas and propagation, Philadelphia, 1986, pp.777-780.

[6] Nakano, H., Nogami, K., Arai, S., Mimaki, H. and Yamauchi, J., "A spiral antenna backed by a conducting plane reflector", IEEE Trans., Vol.AP-34, No.6, 1986, pp.791-796.

[7] Nakano, H., Kerner, S. R. and Alexopoulos, N. G., "The moment method solution for printed antennas of arbitrary configuration", Proc. IEEE International symposium on antennas and propagation, Virginia, 1987, pp.1016-1019.
(Note that a misprint of $k^2=\omega^2\sqrt{\mu_0\varepsilon_0}$ in Eq.(1) should be changed to $k=\omega\sqrt{\mu_0\varepsilon_0}$.)

CHAPTER 4
Spiral Antenna with Two Off-Centre Sources*

4.1. INTRODUCTION

The rotational sense of a circularly polarised wave from a two-wire spiral antenna is basically determined by the winding direction of the two spiral arms. In this chapter, for eliminating the restriction concerning the rotational sense, we introduce a new method of exciting the two-wire spiral antenna from two off-centre sources.

First, consideration is given to the radiation characteristics as the sources are shifted from the centre of the spiral toward the arm ends. The numerical and experimental results reveal that the radiation wave changes from a circularly polarised wave of right-handed sense to one of left-handed sense, by way of an intermediate stage in which a linearly polarised wave is produced.

Secondly, the frequency characteristics of a spiral antenna with fixed off-centre sources are analysed, and the behaviour of the polarisation is discussed, using the analysed current distribution. It will be found that the present spiral antenna realises attractive polarisation characteristics; i.e., one sense of circular polarisation is obtained at a frequency of "f" and the other sense of circular polarisation at a frequency of "2f". This fact leads to the possibility that the spiral be used as a dual-frequency primary feed, having the advantage of the invariance of the phase-centre location.

---

* From a paper (p-18) listed at the end of this monograph.

4.2. RADIATION CHARACTERISTICS AS A FUNCTION OF THE LOCATIONS OF THE TWO SOURCES [1]

Fig.4.1 illustrates the antenna configuration and co-ordinate system. The position vector from the origin to a point of the winding angle ϕ_s on the smooth arm is defined as $\bar{r}=a\phi_s\hat{r}$ (a=spiral constant), as shown in Fig.4.1(b). In the present model, we use a=0.6366 mm/rad, 3.4π rad < ϕ_s < 18π rad, and the radius of the wire conductor ρ=0.2 mm. A straight wire is inserted between the two points which are the beginnings of the smooth arms, and is separated by an infinitesimal gap at the centre. The antenna is excited at two points on the arms, which are symmetrical with respect to the origin, and the two sources are put on the X-axis, as shown in Fig.4.1(c). The location of each source is designated as L_1^+ , L_2^+ , ······· , L_n^+ on the positive X-axis and as L_1^- , L_2^- , ······· , L_n^- on the negative X-axis.

It should be noted that when the spiral is excited from the infinitesimal gap, L_0, the situation is the same as the conventional centre-fed spiral antenna with a single source, which has been described in Chapter 3. The spiral antenna excited from L_0 is used as a reference antenna when we investigate the present spiral antenna with the two sources.

The two sources are assumed to be delta-function generators having the phase relation shown in Fig.4.1(c). In this section, the operating frequency is fixed at 3.0 GHz, where the spiral has an outer circumference of 2.2 λ (λ=wavelength) and a space between the adjacent wires of 1.6×10^{-2} λ. The locations of the two sources are shifted from the centre of the spiral toward the arm end.

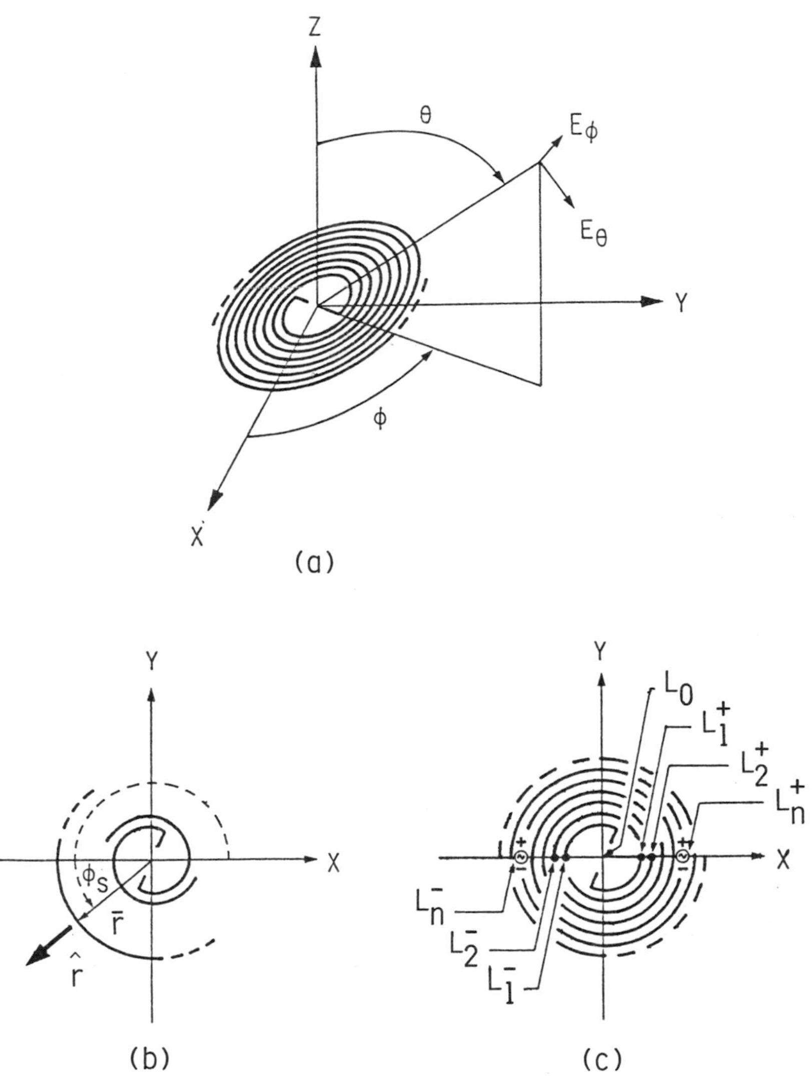

Fig.4.1 Antenna configuration and co-ordinate system.

4.2.1. Active region of the antenna

The radiation mechanism of the present spiral antenna is inferred using the current band theory [2]. Letting two points P and Q lie on the same circle (radius=r) centred with respect to the origin O in Fig.4.2, we give their neighbouring points as P' and Q'. When the space between the neighbouring wires, Δr, is much smaller than r, the arc length QP' along the spiral is approximately equal to πr. Hence, the arc length becomes $\lambda/2$ when r is $\lambda/2\pi$, or when the circumference is λ.

Now, we assume that each wire supports a progressive wave of the current flowing from each source toward outer arm end. Then the difference in the phase of the

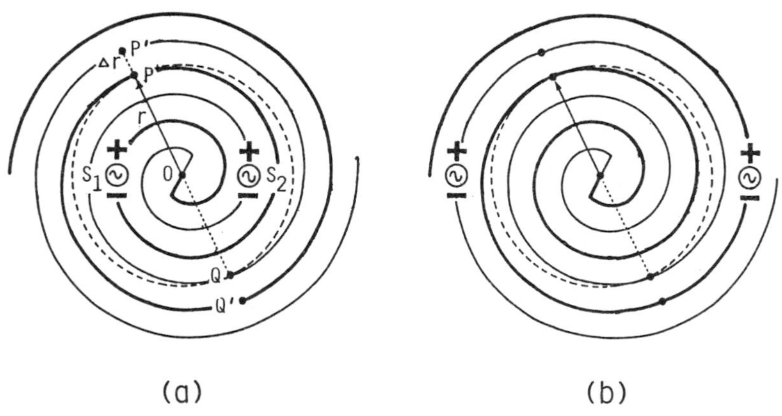

Fig.4.2 Active region of the antenna.

(a) Two sources are inside the one-wavelength circumference.

(b) Two sources are outside the one-wavelength circumference.

one-wavelength circumference

current between the two points P and P' is given by $\pi+(\lambda/2)(2\pi/\lambda)=2\pi$. In other words, the currents at the two points are in-phase. The π rad of the first term is due to the fact that the currents at the two points P and Q are anti-phase owing to the phase of each applied voltage.

The condition for in-phase currents also occurs at the two points Q and Q', which are diametrically opposite to the points P and P' with respect to the origin O, respectively. Thus an active region is generated on an annular ring of turns of one-wavelength mean circumference, from which the strong radiation occurs. We call it the first mode radiation. It is noted that the contribution of the currents flowing toward the origin to the radiation is negligible.

The radiation mechanism mentioned above also holds for the spiral whose two sources are outside the one-wavelength circumference. For the spiral shown in Fig.4.2(b), the currents flowing toward the origin generate the active region on the one-wavelength circumference.

4.2.2. Current distribution

The qualitative explanation of the radiation mechanism has been drawn in the previous section. In this section a rigorous current distribution is determined in order to evaluate the radiation characteristics. The current distribution along the spiral arm is numerically determined by using an integral equation (1.64). It is noted that in the left side of eqn.(1.64) we use only the first term because the magnetic current does not exist, while adding a term due to the effect of the voltage source to the right side [3].

Numerical results except (a) in Fig.4.3 show the current

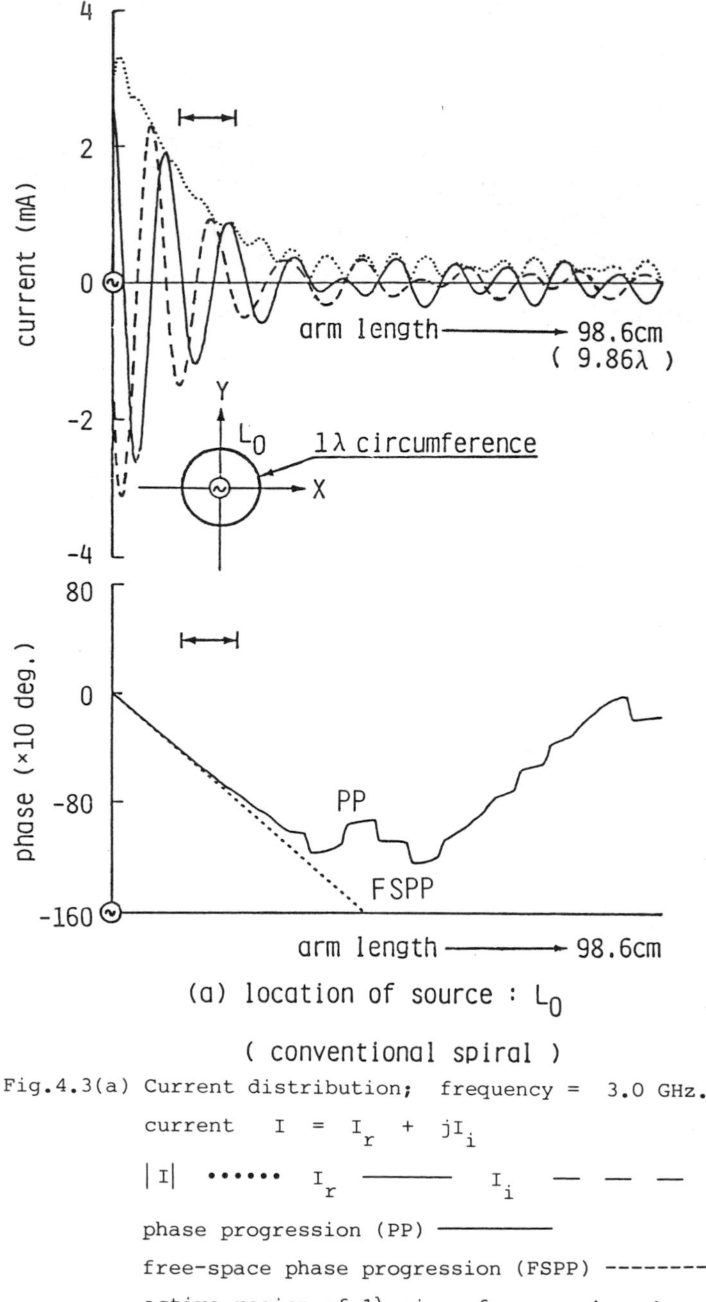

(a) location of source : L_0

(conventional spiral)

Fig.4.3(a) Current distribution; frequency = 3.0 GHz.

current $I = I_r + jI_i$

$|I|$ •••••• I_r ——— I_i — — —

phase progression (PP) ———

free-space phase progression (FSPP) --------

active region of 1λ circumference |←→|

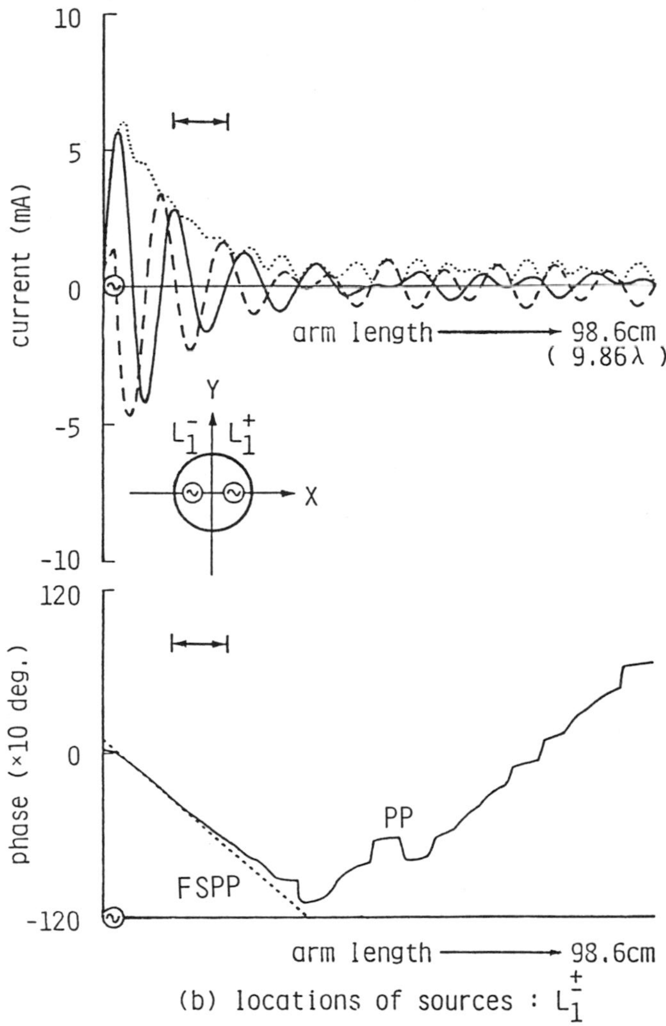

Fig.4.3(b) Current distribution; frequency = 3.0 GHz.

current $I = I_r + jI_i$

$|I|$ •••••• I_r ———— I_i — — —

phase progression (PP) ————

free-space phase progression (FSPP) --------

active region of 1λ circumfernce |←→|

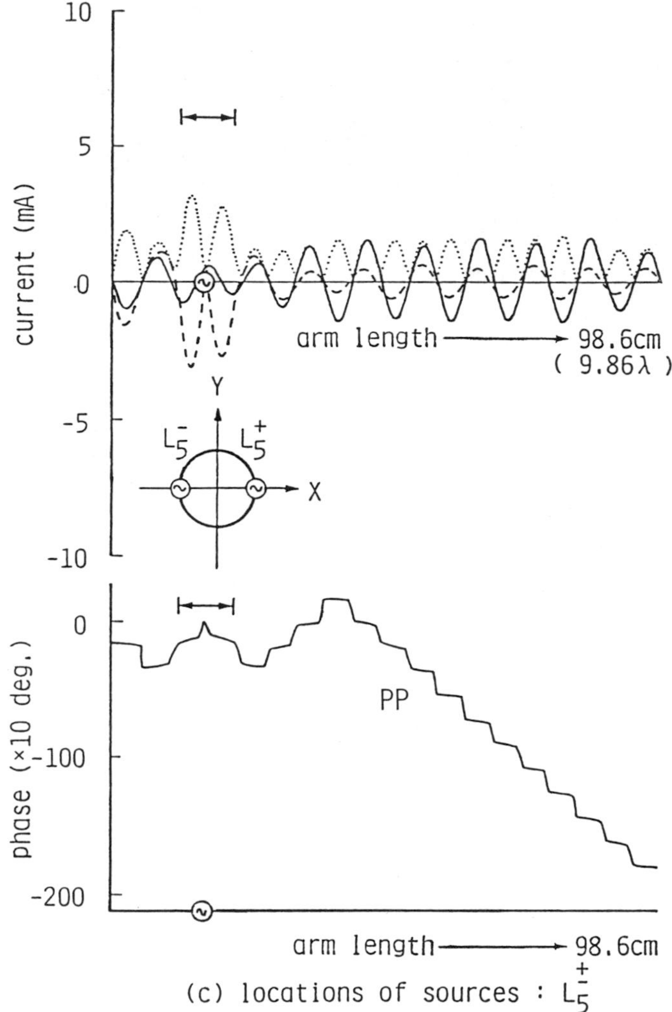

Fig.4.3(c) Current distribution; frequency = 3.0 GHz.

current $I = I_r + jI_i$

$|I|$ ······ I_r ——— I_i — — —

phase progression (PP) ———

active region of 1λ circumference |←→|

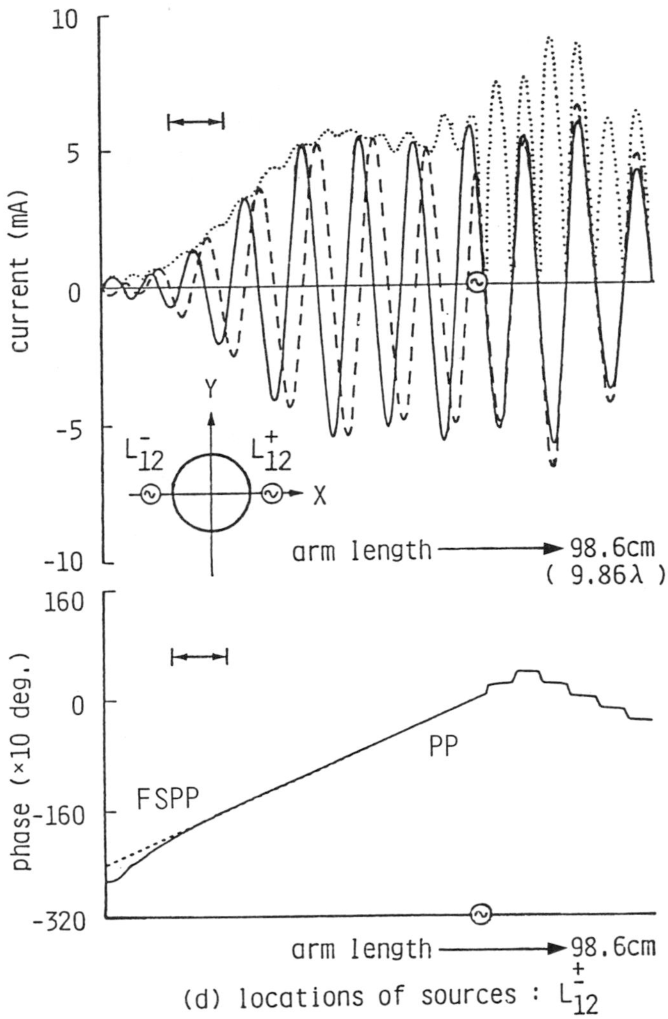

(d) locations of sources : $L_{12}^{\pm}$

Fig.4.3(d) Current distribution; frequency = 3.0 GHz.

current $I = I_r + jI_i$

$|I|$ •••••• I_r ——— I_i — — —

phase progression (PP) ———

free-space phase progression (FSPP) --------

active region of 1λ circumference |←→|

distributions and their phase progressions as the two sources are shifted on the arms. Since the current distribution is symmetrical with respect to the origin in Fig.4.1, half of the current distribution is presented in these results. The inserts illustrate the relative location of the two sources and the one-wavelength circumference. The current distribution of the conventional spiral antenna, in which a single source is located at the origin L_0, is also shown in Fig.4.3(a).

It is worth mentioning that the behaviour of the current distribution shown in Fig.4.3.(b) is similar to that of the conventional spiral. The current smoothly decays over the main region from the source point to the location of the first mode radiation. In contrast, when the two sources are on the one-wavelength circumference, a current standing wave becomes noticeable, as shown in Fig.4.3(c). Consequently, the radiation field is linearly polarised, as will be mentioned in Section 4.2.3.

When the two sources are outside the one-wavelength circumference, as shown in the insert of Fig.4.3(d), we again observe a decaying current in the active region on the one-wavelength circumference. Detailed calculation shows that the rate of decay in the vicinity of the active region is almost the same as that shown in Fig.4.3(b). The rate of decay is about -7 dB/λ. It should be noted that, owing to the difference between the source locations, the phase progressions of Figs.4.3(b) and (d) are opposite to each other in the active regions. In Fig.4.3(b) the travelling current flows from the source point toward the outer arm end. On the other hand, in Fig.4.3(d) the travelling current flows from the source point toward the inner arm end. The difference in the flowing direction between the travelling currents results in the difference in the rotational sense of a circularly polarised wave.

4.2.3. Radiation pattern and axial ratio

Figs.4.4(a)-(d) show the radiation patterns and axial ratios versus the angle θ shown in Fig.4.1(a). These are calculated by using the current distributions illustrated in Figs.4.3.(a)-(d).

When the two sources are inside the one-wavelength circumference, the radiation pattern is almost the same as that of the conventional spiral, as shown in Figs.4.4(a) and (b). This is easily understandable from the similarity between the current distributions illustrated in Figs.4.3(a) and (b). The axial ratio is 1.5 dB on the Z-axis and remains less than 3 dB over an angle range of ±53°. It is noted that the rotational sense of a circularly polarised wave is right-handed on the +Z-axis, which is the same as that of the conventional spiral with a single source located at its centre. The rotational sense of the circularly polarised wave is confirmed by using a balanced helical antenna [4], which will be discussed in Section 6.5. When the two sources are on the one-wavelength circumference as illustrated in the insert of Fig.4.4(c), the remarkable difference between the levels of the E_ϕ- and E_θ- components is observed, indicating that the radiation field is linearly polarised. The axial ratio is 23.6 dB at $\theta=0°$ and is more than 10 dB over an angle range of ±65°.

Fig.4.4(d) is the case where the two sources are outside the one-wavelength circumference. We observe the close agreement in the levels of the E_ϕ- and E_θ- components over a wide angle around the Z-axis. The axial ratio is 2.7 dB at $\theta=0°$ and is within 3 dB over an angle range of ±45°. Theoretical and experimental results reveal that the rotational sense of a circularly polarised wave is left-handed on the +Z-axis, which is opposite to the rotational sense for the two sources being inside the

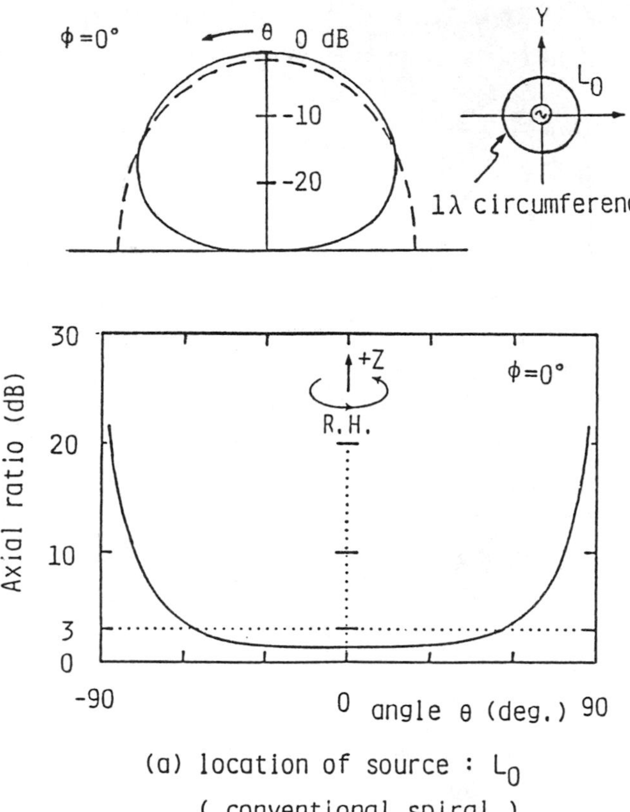

Fig.4.4(a) Radiation pattern and axial ratio versus angle θ; frequency=3.0 GHz.

radiation pattern: theoretical $\begin{cases} E_\phi & \text{— —} \\ E_\theta & \text{——} \end{cases}$

axial ratio: theoretical ——

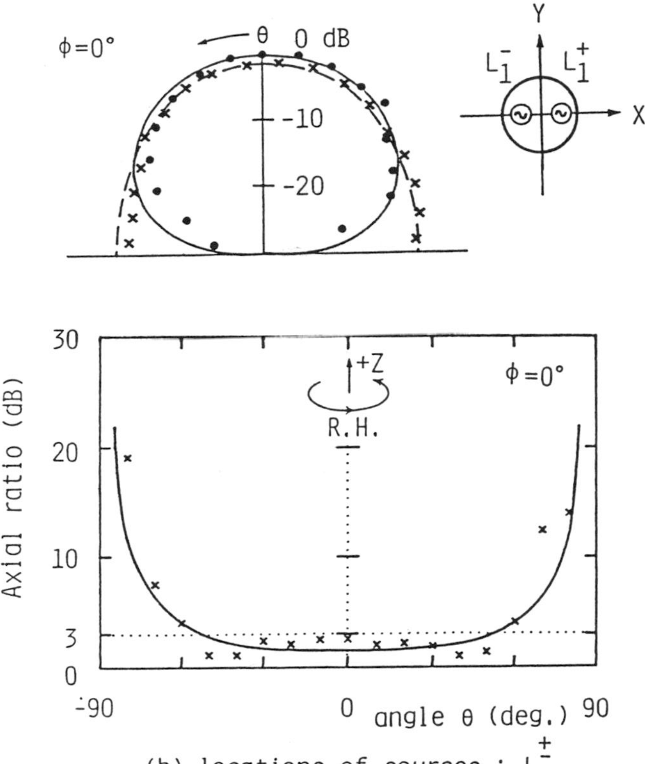

(b) locations of sources : $L_1^{\pm}$

Fig.4.4(b) Radiation pattern and axial ratio versus angle θ; frequency=3.0 GHz.

radiation pattern: theoretical { E_ϕ — — ; E_θ ——— }

experimental { E_ϕ xxxx ; E_θ •••• }

axial ratio: theoretical ———

experimental xxxxx

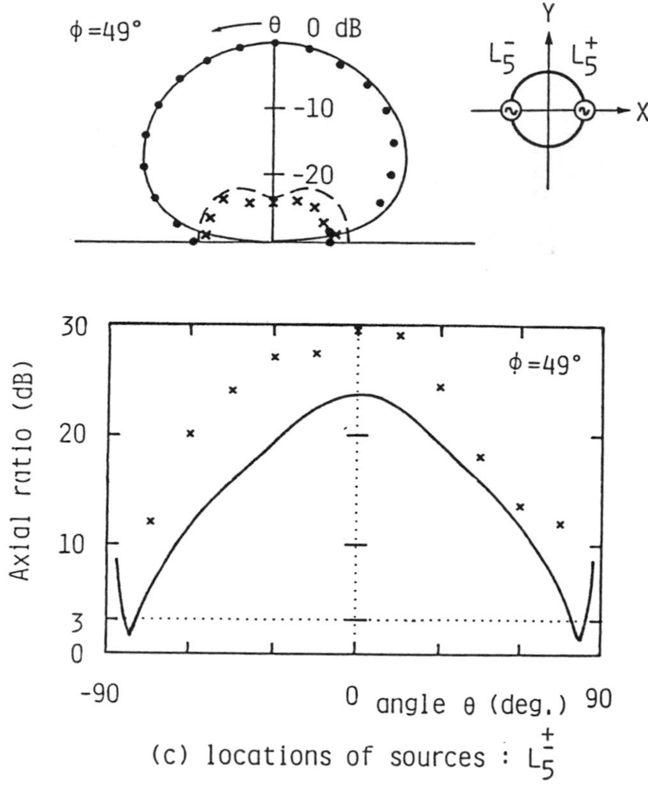

(c) locations of sources : $L_5^{\pm}$

Fig.4.4(c) Radiation pattern and axial ratio versus angle θ; frequency=3.0 GHz.

radiation pattern: theoretical $\begin{cases} E_\phi & — — \\ E_\theta & —— \end{cases}$

experimental $\begin{cases} E_\phi & \times\times\times\times \\ E_\theta & \bullet\bullet\bullet\bullet \end{cases}$

axial ratio: theoretical ——

experimental ×××××

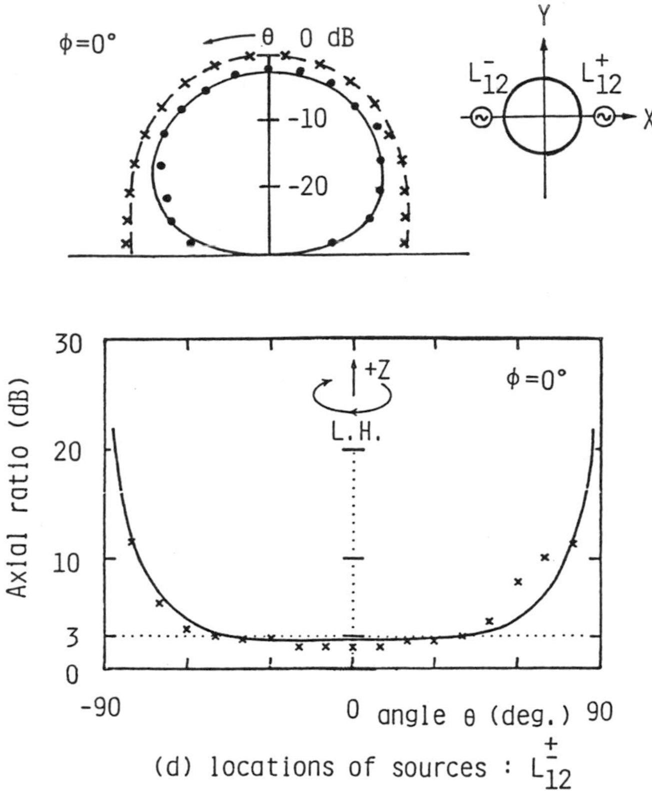

(d) locations of sources : $L_{12}^{\pm}$

Fig.4.4(d) Radiation pattern and axial ratio versus angle θ; frequency=3.0 GHz.

radiation pattern: theoretical $\begin{cases} E_\phi & — — \\ E_\theta & ——— \end{cases}$

experimental $\begin{cases} E_\phi & \times\times\times\times \\ E_\theta & \bullet\bullet\bullet\bullet \end{cases}$

axial ratio: theoretical ———

experimental ×××××

one-wavelength circumference.

From the results mentioned above it can be said that the rotational sense of the radiation field depends on the relative location of the two sources and the active region of the one-wavelength circumference. Since the relative location is also changed by varying the operating frequency, it is expected that a spiral antenna with two "fixed" sources becomes a radiator of circular polarisation of right-handed sense at a frequency of "f" and one of circular polarisation of left-handed sense at a frequency of "2f".

As the operating frequency is increased, the spiral becomes electrically large enough to contain a three-wavelength circumference. It should be noted that on the three-wavelength circumference the currents on the adjacent arms become again in-phase ($\pi+(3\lambda/2)(2\pi/\lambda) = 4\pi$) and generate an active region of the so-called third mode radiation. In the following section we investigate the radiation characteristics when the operating frequency is changed.

4.3. FREQUENCY CHARACTERISTICS OF A SPIRAL ANTENNA WITH TWO FIXED SOURCES

When the two sources are fixed at source locations of $L_5^{\pm}$, the spiral antenna radiates a linearly polarised wave at a frequency of 3.0 GHz, as shown in Fig.4.4(c). There is the possibility of a spiral antenna radiating a circularly polarised wave of right-handed sense on the +Z-axis at frequencies lower than 3.0 GHz and radiating a circularly polarised wave of left-handed sense at frequencies higher than 3.0 GHz. From this viewpoint, an analysis is continued over a frequency range of 2.0 GHz to 5.0 GHz, under the condition that the locations of the two sources are fixed at $L_5^{\pm}$.

Calculation at frequencies lower than 3.0 GHz shows that the current distribution and radiation characteristics are, respectively, similar to those described in Fig.4.3(b) and Fig.4.4(b), where the two sources are inside the one-wavelength circumference. For example, the characteristics at 2.0 GHz are similar to those obtained in the case where the locations of the two sources are $L_3^{\pm}$ at 3.0 GHz.

On the other hand, even though the frequency is increased from 3.0 GHz to 5.0 GHz and the one-wavelength circumference moves inside, the current distribution and radiation characteristics do not show resemblance to those described in Fig.4.3(d) and Fig.4.4(d), respectively. Fig.4.5(a) shows the current distribution at 5.0 GHz (λ=0.06 m). In this case the outer circumference and the space between the adjacent wires are 3.7 λ and 2.7×10^{-2} λ, respectively, and the locations of the two sources, $L_5^{\pm}$, are outside the one-wavelength circumference, as illustrated in the insert of Fig.4.5(a). We observe a travelling current in the active region of the one-wavelength circumference, as in the case of Fig.4.3(d). However, the current which flows toward the outer arm end is different from that in Fig.4.3(d). It decays over the active region on the three-wavelength circumference, generating the third mode radiation.

The radiation pattern which is formed by the radiation of the first and third modes is shown in Fig.4.5(b). The pattern of the E_ϕ component is broader, while that of the E_θ component is narrower, compared with those in Fig.4.4(d) formed only by the first mode radiation. The half-power beamwidth of the E_θ component in the ϕ=0° plane is decreased from ±37° to ±24° due to the existence of the third mode radiation. Because of the difference between the directions of the flowing currents on the two active

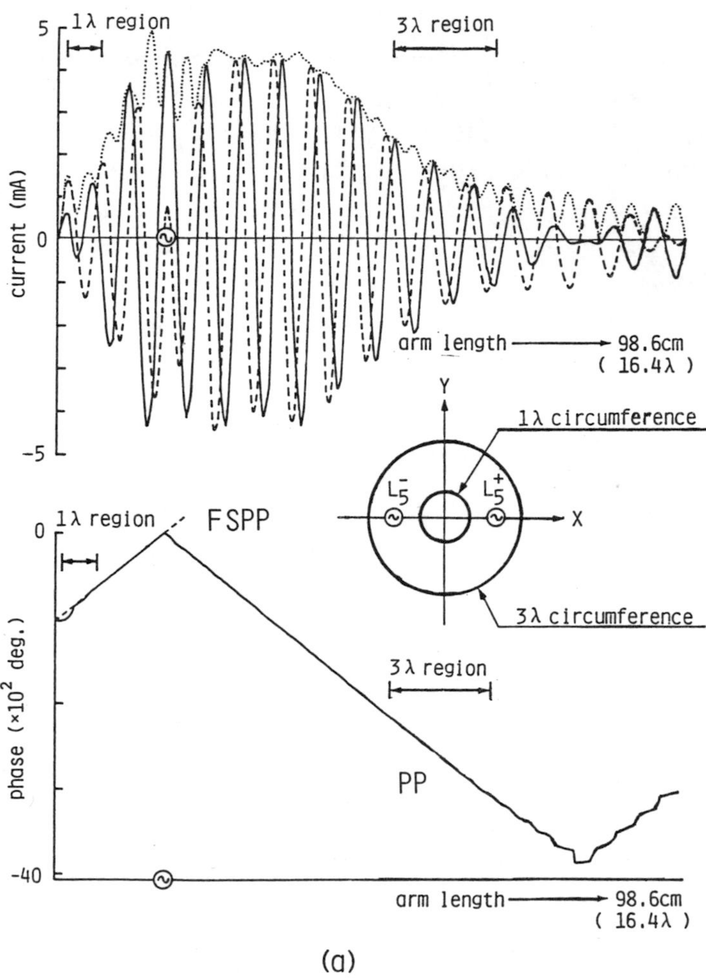

Fig.4.5(a) Current distribution; frequency = 5.0 GHz.

current $I = I_r + jI_i$

$|I|$ ······· I_r ———— I_i — — —

phase progression (PP) ————

free-space phase progression (FSPP) ---------

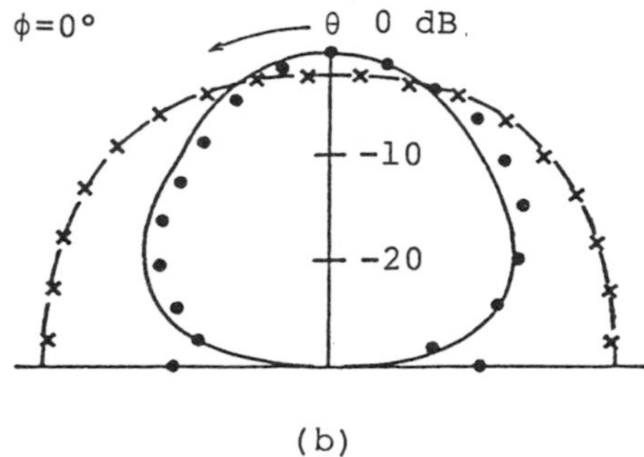

(b)

Fig.4.5(b) Radiation pattern; frequency = 5.0 GHz.

theoretical { E_ϕ — — , E_θ ——— } experimental { E_ϕ ××× , E_θ ••• }

regions, the spiral antenna simultaneously radiates two kinds of circularly polarised waves whose rotational senses are right-handed and left-handed, but yet it does not radiate a linearly polarised wave on the Z-axis. This is interpreted as follows: a circularly polarised wave by the first mode radiation is dominant around the Z-axis, since the third mode radiation makes a conical beam which has a minimum on the Z-axis.

Fig.4.6 shows the axial ratio as a function of frequency. As expected, a frequency of 3.0 GHz is regarded as a turning frequency at which the rotational sense of the radiation field is changed from a right-handed sense to a left-handed sense. An important point is that the spiral radiates a circularly polarised wave of right-handed sense at a frequency of "f" and radiates one of left-handed sense at a frequency of "2f". For instance, this relation is held at frequencies of 2.5 GHz and 5.0 GHz.

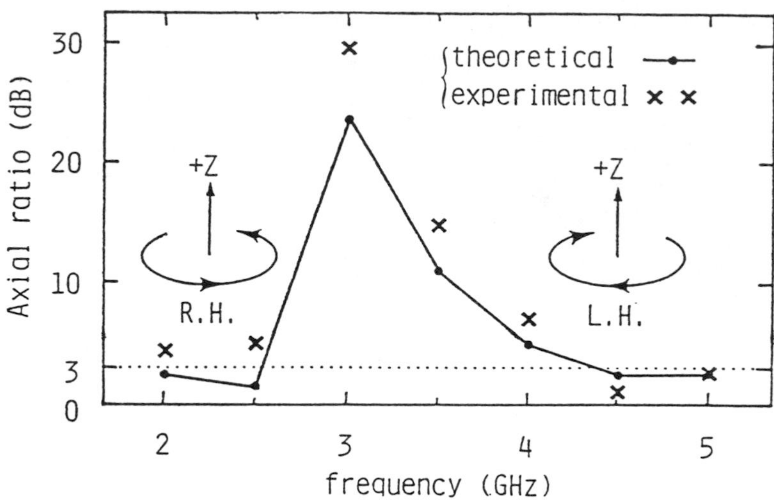

Fig.4.6 Axial ratio versus frequency.

REFERENCES TO CHAPTER 4

[1] Nakano, H., Hirose, K. and Yamauchi, J., "Spiral antennna with two off-center sources", Trans. IECE of Japan, Vol.E67, 6, 1984, pp.309-314.

[2] Kaiser, J.A., "The archimedean two-wire spiral antenna", IRE Trans. Antennas and Propag., AP-8 , 3,pp.312-323(May 1960).

[3] Nakano, H., "The integral equations for a system composed of many arbitrarily bent wires", Trans. IECE of Japan, Vol.E65, 6, 1982, pp.303-309.

[4] Yamauchi, J. and Nakano, H., "Axial ratio of balanced helical antenna and ellipticity measurement of incident wave", Electron.Lett., 17, pp.365-366 (May 1981).

APPENDIX

SPIRAL ANTENNA WITH ASYMMETRICAL ARMS*

The purpose of this appendix is to theoretically reveal the radiation characteristics of an asymmetrical spiral antenna with two off-centre sources.

Fig.A.4.1 shows the antenna configuration and the excitation modes. In the present spiral, the outer circumference is taken to be 1.5 λ, and the spiral arm length prior to being truncated is 4.73 λ. The spiral is excited at two symmetrical points with respect to the origin. These two points are located near the origin, and the distance between the two points is 0.125 λ. The excitation modes shown in Figs.A.4.1(b) and (c) are called "anti-phase excitation" and "in-phase excitation", respectively.

The theoretical current distributions for an asymmetrical spiral antenna with a truncation amount of $\ell=\lambda/4$ are shown in Fig.A.4.2. It is found that for the anti-phase excitation the currents gradually decay owing to radiation. The current distribution has a form similar to that of a conventional centre-fed spiral antenna with symmetrical arms. Since the winding sense of the spiral as viewed from the -Z side is right-handed as shown in Fig.A.4.1(a), the travelling currents toward the arm ends radiate a circularly polarised wave of right-handed sense on the +Z-axis.

The theoretical current distribution for the in-phase excitation shows a standing wave form, as shown in Fig.A.4.2(b). It should be noted that the standing-wave ratio of the current reduces toward unity, as the point of observation is moved away from the arm end toward the

---

* From a paper (p-22) listed at the end of this monograph.

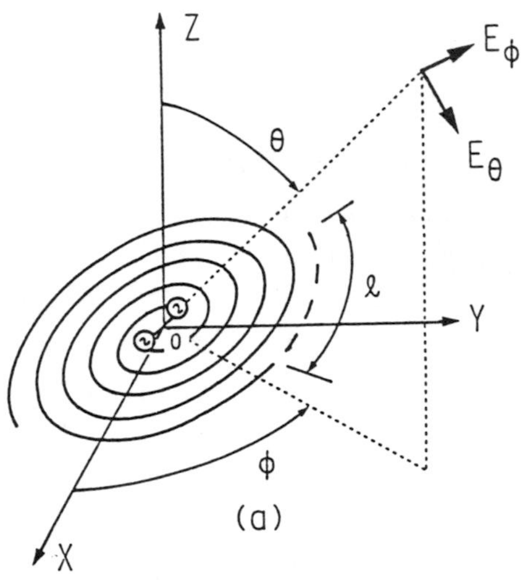

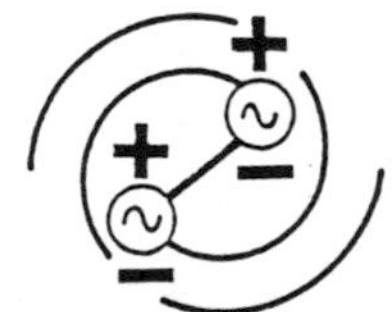

(b) anti-phase excitation

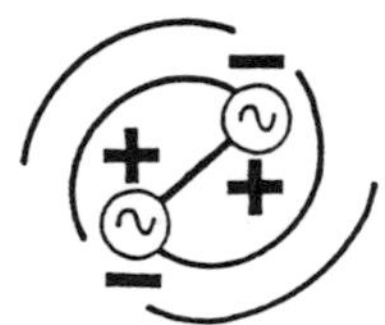

(c) in-phase excitation

Fig.A.4.1 Antenna configuration and excitation modes.

(a) Antenna configuration.

(b) Anti-phase excitation.

(c) In-phase excitation.

spiral constant $a=5.83\times10^{-3}\lambda$ (m/rad)

wire radius$=2.33\times10^{-3}\lambda$ (m)

maximum winding angle $\phi_s=41.4$ rad

origin. This behaviour can be explained in terms of the interference of two currents. One is a travelling current from the source toward the arm end with little attenuation, and the other is reflected current from the arm end toward the source, decaying owing to radiation. The reflected current radiates a circularly polarised wave of left-handed sense on the +Z-axis, which is the opposite of the sense obtained for the anti-phase excitation. Thus, selection of either sense of circular polarisation is realised by changing the excitation mode.

Fig.A.4.3 shows the theoretical and experimental radiation patterns. It is seen that the broad axial beam is radiated for each excitation mode. The axial ratios for the anti-phase excitation and in-phase excitation are 1.3 dB and 1.8 dB, respectively. It is worth mentioning that both senses of a circularly polarised wave are obtained with the radiation pattern being nearly unchanged.

So far, the study has been restricted to the case where the truncation amount on the one arm is $\lambda/4$. It is significant to examine how the change in the truncation amount ℓ affects the radiation characteristics.

Fig.A.4.4 shows the axial ratio on the Z-axis as a function of the truncation amount. The axial ratio for the anti-phase excitation is almost independent of the truncation amount. Axial ratios of less than 2.0 dB are obtained in the direction of the Z-axis. The direction of the maximum radiation remains on the Z-axis even when the truncation amount is $\lambda/2$.

For the in-phase excitation, there exists a truncation range in which a circularly polarised wave is radiated. Although axial ratios of less than 3.0 dB are obtained over a relatively wide truncation range, the direction of the maximum radiation tends to deviate from the Z-axis as the truncation amount is increased or decreased from $\lambda/4$.

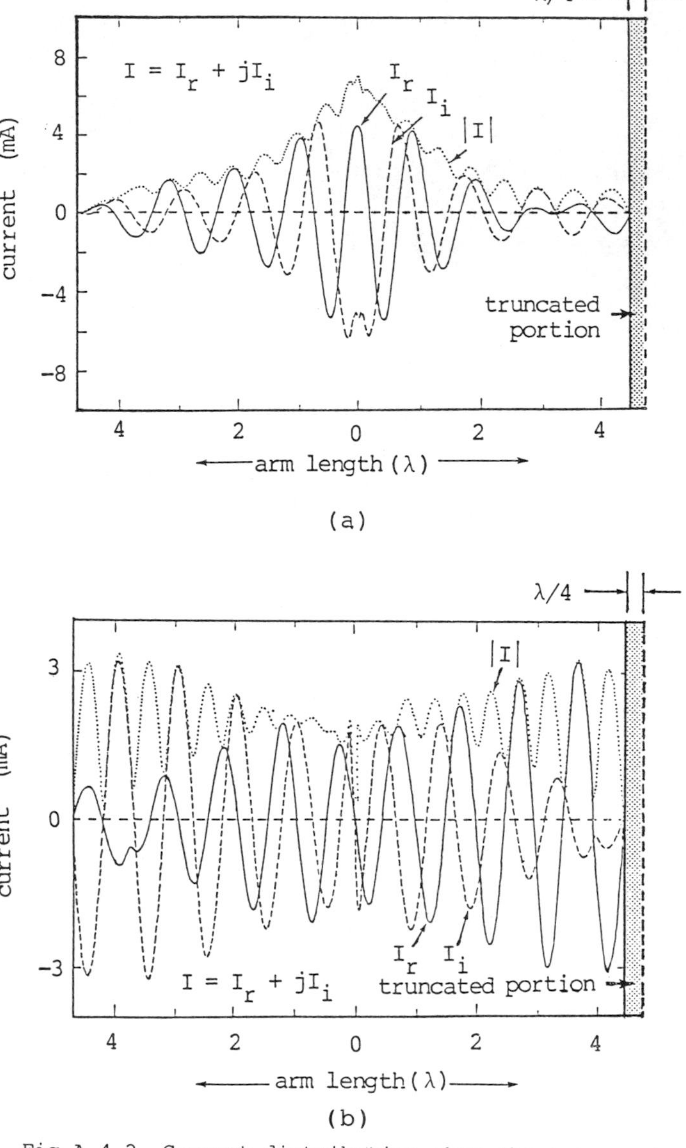

Fig.A.4.2 Current distribution when the one arm is truncated by $\lambda/4$.

(a) Anti-phase excitation.

(b) In-phase excitation.

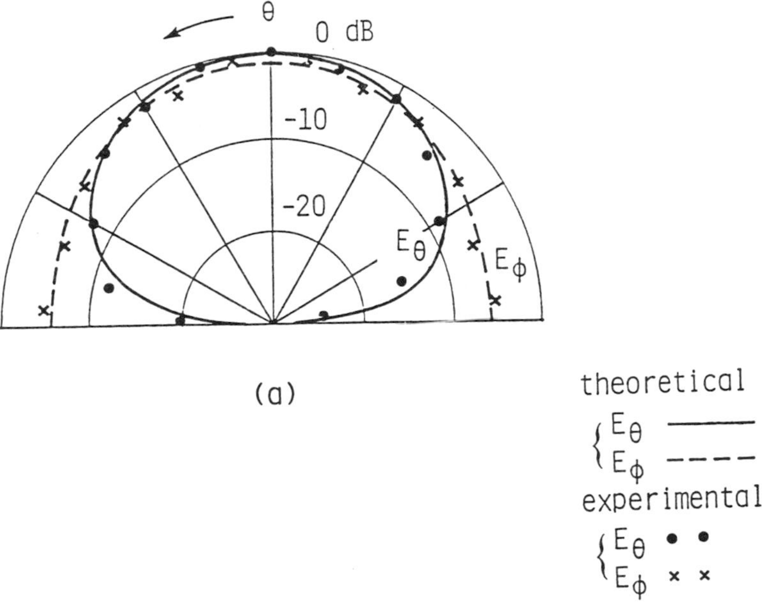

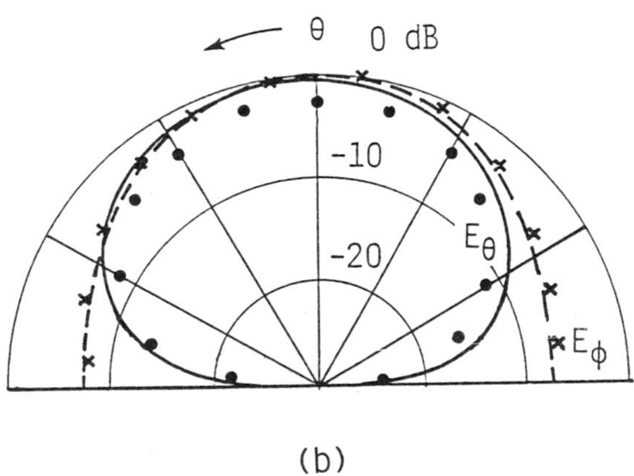

(b)

Fig.A.4.3 Radiation pattern when the one arm is truncated by $\lambda/4$.

(a) Anti-phase excitation.

(b) In-phase excitation.

The sense selection for a tolerable deviation angle within 6° is possible in a truncation range of $2\lambda/16$ to $5\lambda/16$, as shown in Fig.A.4.4.

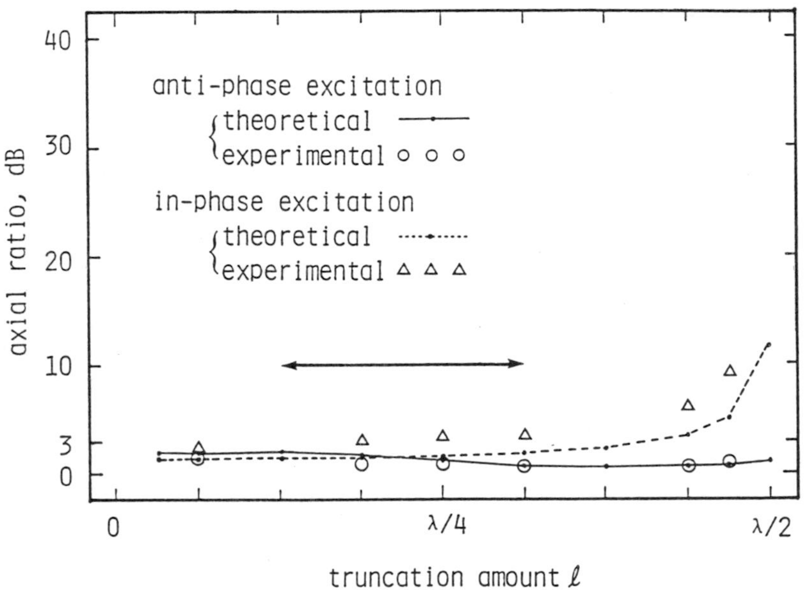

Fig.A.4.4 Axial ratio as a function of truncation amount of the one arm.

⟷ A region where the deviation angle of the maximum radiation from the Z-axis is within 6°.

CHAPTER 5
Polarisation Diversity of Archimedean Spiral Antenna*

5.1. INTRODUCTION

This chapter describes an Archimedean spiral antenna consisting of driven and parasitic elements on the same plane(SAP), and is aimed at determining the effects of the parasitic element on the radiation characteristics. The current distributions on both the driven and the parasitic elements are determined by solving a pair of integral equations, and compared with those of a two-wire spiral antenna and a four-wire spiral antenna with four-terminal feed. It will be demonstrated that the SAP operates as a radiation element of linear polarisation at low frequencies and of circular polarisation at high frequencies.

In Section 5.4, for obtaining polarisation diversity at a given frequency, we consider the case where all the four arms are fed. It will be quantitatively clarified that a circularly polarised wave of the right-handed or left-handed rotational sense can be obtained for a broad range of angles by changing the excitation phase [1].

5.2. ANTENNA GEOMETRY AND NUMERICAL METHOD [2]

Fiq.5.1 shows the configuration and co-ordinate system for an SAP, i.e., a spiral antenna with a parasitic

---

*From papers (p-13), (p-17), (c-9) and (c-15) listed at the end of this monograph.

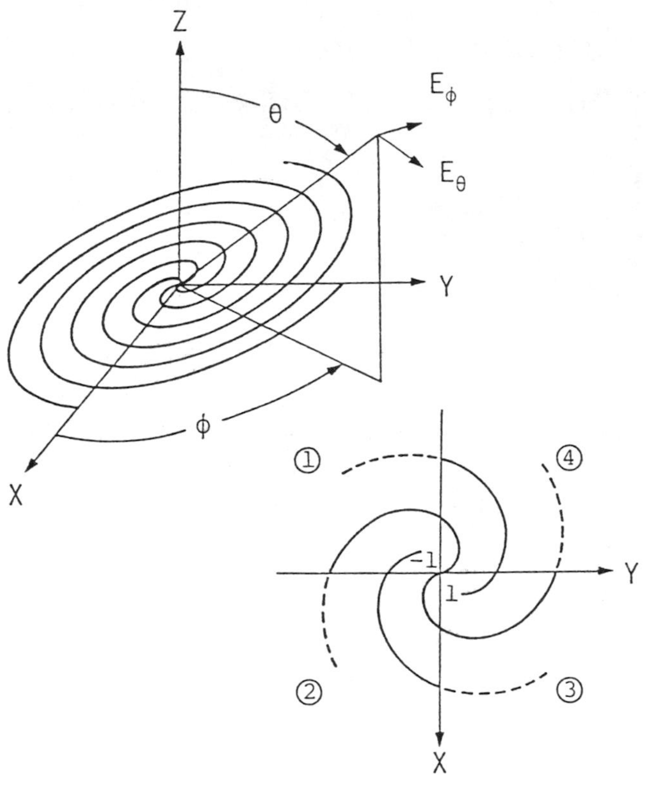

Fig.5.1 Configuration and co-ordinate system of SAP.

element consisting of two arms joined at the origin. The parasitic arms are rotated by 90 degrees from the driven arms. The SAP, therefore, corresponds to a four-wire spiral antenna in which two of the four arms (arms 2 and 4) are not excited.

We use a frequency range of 3 GHz to 6 GHz, and choose the antenna parameters as follows: spiral constant is a=1.5 mm/rad and the number of turns is t=4, resulting in an antenna peripheral length (outer circumference) of about 24 cm. Depending on the operating frequency, the peripheral length varies from 2.4 λ to 4.8 λ (λ=free-space wavelength). The wire radius is taken to be 0.3 mm throughout all the sections.

An analysis is made for the case of exciting arms 1 and 3 with an applied voltage of V_0 = 1 volt. From the symmetry of the currents on arms 1 and 3 and that of the currents on arms 2 and 4, the following integral equations are obtained for the currents $I_1(s')$ and $I_2(s')$, which flow on the driven arm 1 and the parasitic arm 2, respectively:

$$\int_0^L I_1(s')(\pi_{11}-\pi_{13})ds' + \int_0^L I_2(s')(\pi_{12}-\pi_{14})ds'$$

$$= B\cos\beta s - \frac{jV_0}{2Z_0}\sin\beta s ,$$

<u>s : on arms 1 and 3</u>

$$\int_0^L I_1(s')(\pi_{21}-\pi_{23})ds' + \int_0^L I_2(s')(\pi_{22}-\pi_{24})ds'$$

$$= B'\cos\beta s ,$$

<u>s : on arms 2 and 4</u>

(5.1)

where

$$\pi_{\ell,i} = G_{\ell,i}(s,s')\hat{s}_\ell \cdot \hat{s}'_i - \int_0^s \left[\frac{\partial G_{\ell,i}(\xi,s')}{\partial s'} + \frac{\partial}{\partial \xi} \{ G_{\ell,i}(\xi,s')\hat{\xi}_\ell \cdot \hat{s}'_i \} \right] \cdot \cos\beta(s-\xi)d\xi$$

$$\begin{cases} \ell=1,2 \\ i=1,2,3,4 \end{cases} \tag{5.2}$$

in which $G_{\ell,i}(s,s')$ is free-space Green's function; $\hat{s}_\ell$, $\hat{s}'_\ell$, and $\hat{\xi}_\ell$ are the tangential unit vectors at the points whose distances measured along the ℓ-th arm from the origin are s, s' and ξ, respectively; L is the antenna arm length; Z_0 is the intrinsic impedance of free-space; $\beta=2\pi/\lambda$; and B and B' are unknown constants which are determined by using the condition that the currents on the arm ends are zero.

The radiation field at an observation point (R, θ, ϕ) is given by

$$\overline{E}(R,\theta,\phi) = -j\omega\mu \frac{e^{-j\beta R}}{4\pi R} \sum_{\ell=1}^{4} \int_0^L \hat{R}\times\{ I_\ell(s')\hat{s}'_\ell \times \hat{R}\} e^{j\beta R_\ell} ds' , \tag{5.3}$$

where

$$R_\ell = a\phi' \sin\theta \cos(\phi'_\ell - \phi) \quad , \tag{5.4}$$

$$[\ \phi'_\ell = \phi' + \frac{\ell-1}{2}\pi \qquad \ell=1,2,3,4\ ]$$

$\hat{R}$ is the unit vector oriented from the origin to the observation point.

5.3. NUMERICAL AND EXPERIMENTAL RESULTS [2]

5.3.1. The first mode radiation and the third mode radiation

Figs.5.2(a)-(d) show the amplitudes of the current distributions when the frequency is changed. At 3 GHz and 4 GHz, standing waves are clearly observed on both the driven and the parasitic arms, while at 5 GHz and 6 GHz the standing waves decrease and the travelling waves with attenuation are dominant.

The phase progressions of the currents on both the driven and the parasitic arms are added to the result at 6 GHz [see Fig.5.2(e)]. They show that the current induced on the parasitic arm 2 has its phase 90 degrees ahead of that on the driven arm 1. Although not shown in the Figure, the phase difference between the currents on the driven and parasitic arms is about 90 degrees at other frequencies. In other words, the phase relation of the currents for the SAP corresponds to that for a four-wire spiral antenna excited by mode 3 [see Fig.5.3(b) and refer to Section 5.4 for defining the mode].

In Fig.5.2(d), the current amplitudes for two cases are also plotted; one is for the case where the parasitic arms are removed, and the remaining arms are excited as a

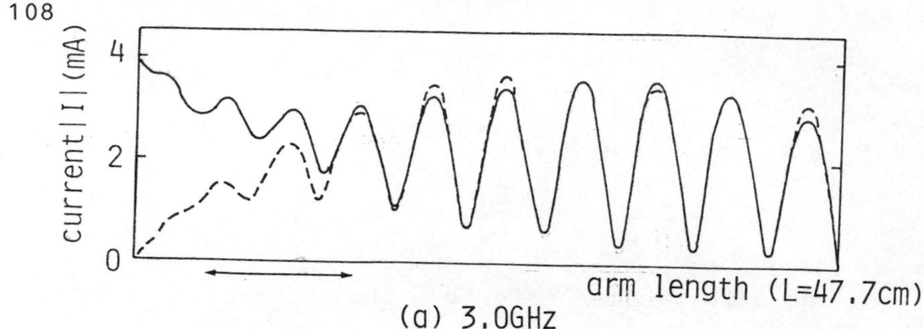

(a) 3.0GHz

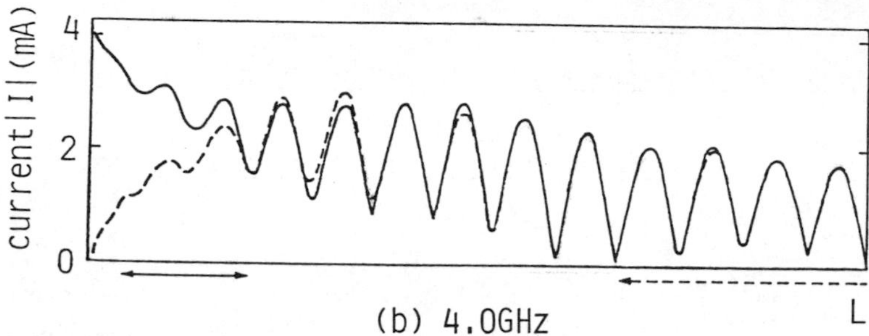

(b) 4.0GHz

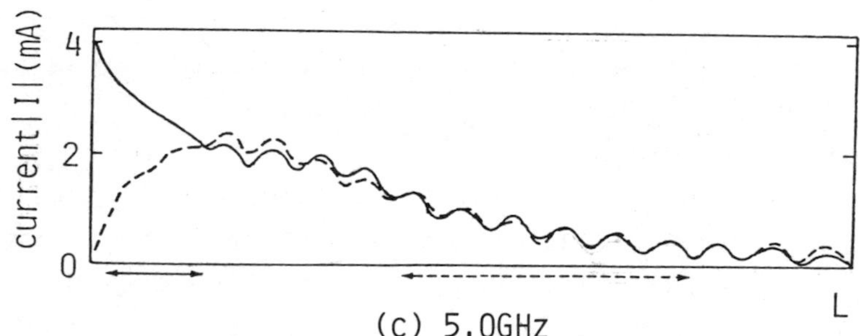

(c) 5.0GHz

Fig.5.2(a),(b),(c) Frequency characteristics of current distribution.

——— current amplitude on driven spiral arm

– – – – – current amplitude on parasitic spiral arm

←——→ location of the first mode radiation

← – – – → location of the third mode radiation

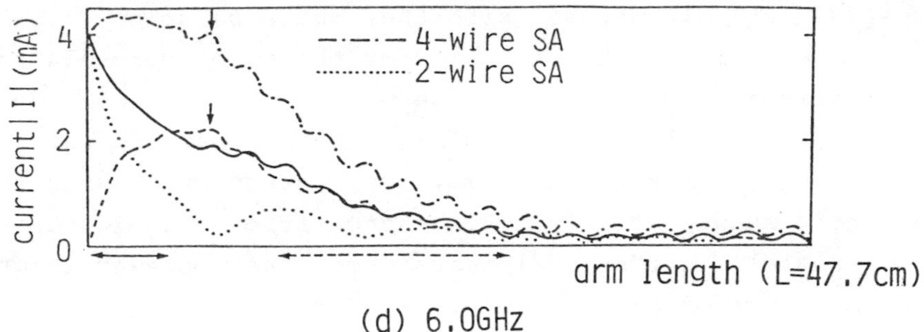

(d) 6.0GHz

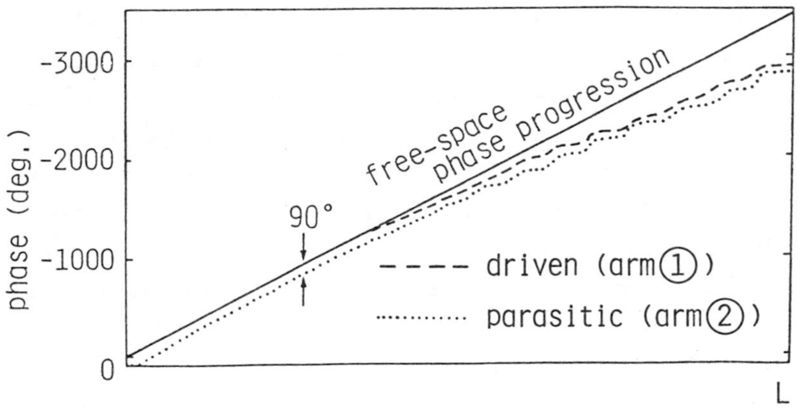

(e) phase progression ; 6.0GHz

Fig.5.2(d),(e) Frequency characteristics of current distribution.

——— current amplitude on driven spiral arm

– – – – – current amplitude on parasitic spiral arm

←——→ location of the first mode radiation

← – – – → location of the third mode radiation

two-wire spiral antenna with the phase relation (1, -1) [see Fig.5.3(a)]; the other is for the case where all the four arms are excited with the phase relation of mode 3 (1, j, -1, -j) [see Fig.5.3(b)]. The behaviour of the radiation for each case may be interpreted with the currents shown in Fig.5.3, where arrows indicate the differential current elements on the one-wavelength circumference of each antenna.

In the two-wire spiral antenna shown in Fig.5.3(a), the currents on the neighbouring arms 1 and 3 are in-phase on the one-wavelength circumference and cause strong radiation, which is called "the first mode radiation".

Under the excitation of mode 3 in the four-wire spiral antenna shown in Fig.5.3(b), the currents on arms 1 and 3 are in-phase on the one-wavelength circumference, but the currents on arms 2 and 4 are out of phase with respect to those on arms 1 and 3. Since the currents on the four arms have almost the same amplitude, the radiation from these currents is extremely small on the Z-axis. It should be noted, however, that strong radiation occurs from the currents on the region whose circumference is three wavelengths, because the currents are in-phase on this region. The radiation is called "the third mode radiation".

In the SAP shown in Fig.5.3(c), the phase relation of the currents corresponds to that under the excitation of mode 3 in the four-wire spiral antenna. This has been described in Fig.5.2(e), together with Fig.5.2(d), where it is obvious that, over the region from the feed point to the one-wavelength circumference, the currents on the parasitic arms 2 and 4 are smaller than those on the driven arms 1 and 3. Therefore, the currents on the driven arms are not cancelled by those on the parasitic arms, and the residual portion contributes to radiation (the first mode radiation). This situation is unlike the

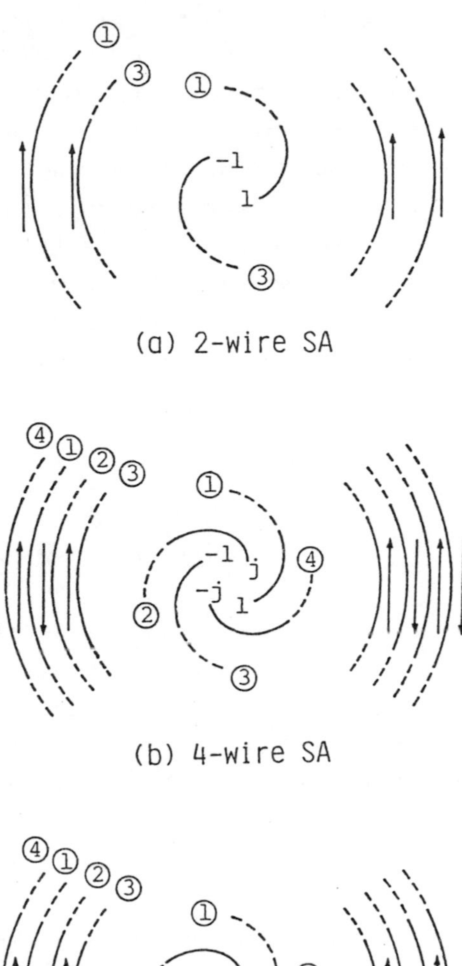

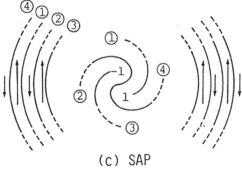

Fig.5.3 Behaviour of the differential current element in the first mode radiation.

previous case of the four-wire spiral antenna. However, as soon as the currents travelling toward the arm ends become equal in their amplitudes, the SAP behaves as the four-wire spiral antenna and generates the third mode radiation. This is understood from the fact that the currents on the SAP and the four-wire spiral antenna start decreasing at almost the same positions [see two arrows ↓ in Fig.5.2(d)]. In summary, the SAP simultaneously generates the first mode radiation and the third mode radiation at the high frequencies which allow the SAP to be large enough to contain the three-wavelength circumference.

5.3.2. Behaviour of reflected currents

Again, we refer to Figs.5.2(a) and (b) and interpret the behaviour of the current distribution. As the frequency is decreased from 6 GHz, the region where the three-wavelength circumference is located on the SAP moves toward the outer arms, and finally it diminishes. Thus the currents after passing the one-wavelength circumference on the driven and parasitic arms arrive at the arm ends without attenuation, and they are reflected there. As a result, the standing wave is observed in the current distributions on both arms.

The radiation influenced by the reflected currents continuing to travel toward the inner region of the SAP, which no longer contains the region of the three-wavelength circumference, is explained as follows : as the reflected currents reach the region of the one-wavelength circumference, they become in-phase and generate strong radiation. This radiation due to the reflected currents is circularly polarised with the rotational sense which is opposite to that of the first mode radiation due to the original travelling-currents flowing out of the feed points or the inner region toward the arm ends. Therefore, the combined radiation is linearly polarised [see Section 5.3.3].

5.3.3. Radiation characteristics

Fig.5.4 shows the input impedance of the SAP, together with that of the two-wire spiral antenna shown in

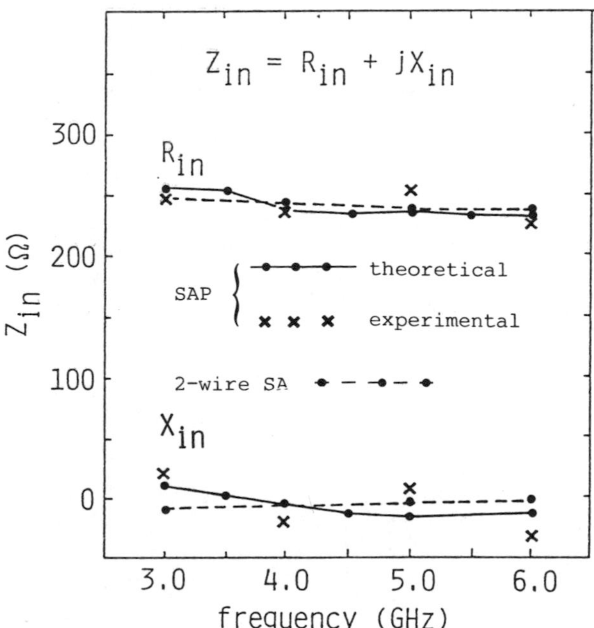

Fig.5.4 Input impedance.

Fig.5.3(a). The two-wire spiral antenna has an impedance which is constant regardless of the change in the frequency. This also holds in the case of the SAP. The impedance is almost purely resistive over a frequency range of 3 to 6 GHz. It is worth mentioning that, despite the standing wave in the current distribution at low frequencies [Figs.5.2(a),(b)], the constant impedance characteristics are maintained. This is because the reflected current on and near the region of the one-wavelength circumference decays due to the first mode radiation and does not affect the feed point.

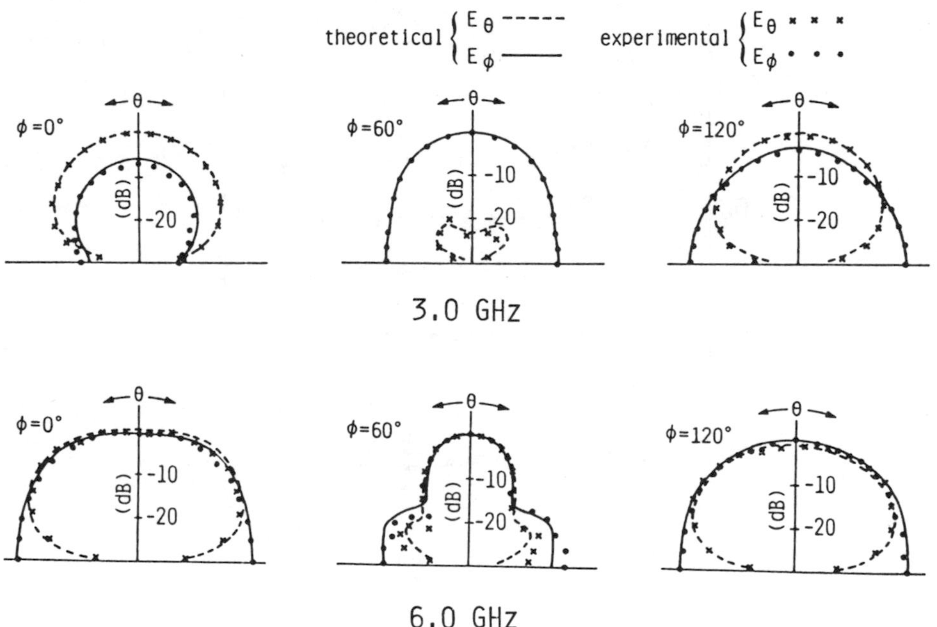

Fig.5.5 Radiation patterns.

Fig.5.5 shows typical radiation patterns of the SAP. At 3 GHz the radiation is linearly polarised with a polarisation plane of about ϕ=150°(=60°+90°), while at 6 GHz the radiation is circularly polarised. As described in Section 5.3.1, the radiation pattern at 6 GHz is the combined one of the first mode radiation and the third mode radiation, the former and the latter being a broad pattern with its maximum on the Z-axis and a two-peak

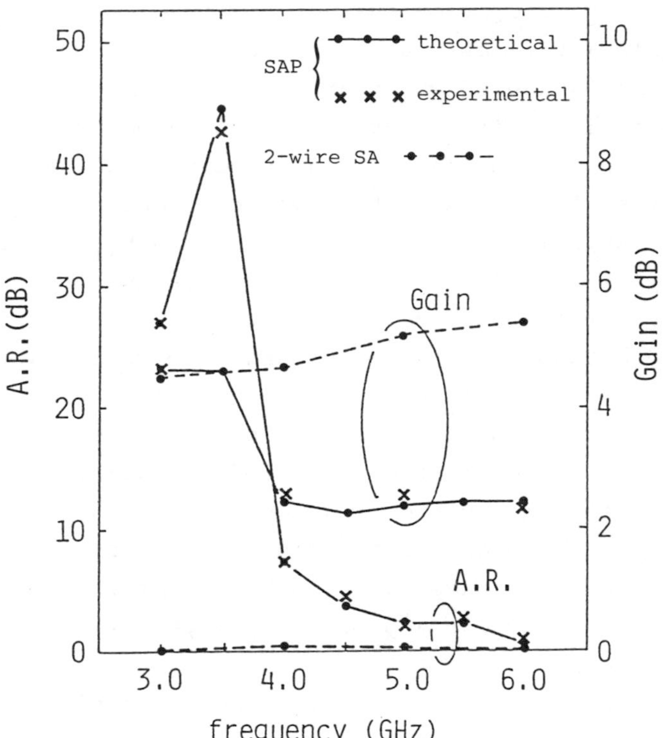

Fig.5.6 Gain and axial ratio.

pattern in which the radiation on the Z-axis is extremely small, respectively.

The gain and axial ratio versus frequency are shown in Fig.5.6. The gain is about 4.5 dB at the lower frequencies and almost equal to the value for the two-wire spiral antenna, although it decreases to about 2.5 dB at higher frequencies.

5.4. POLARISATION DIVERSITY BY FOUR DRIVEN ARMS [2]

So far the radiation characteristics of the SAP have been calculated as a function of frequency. The calculations have shown that the SAP has a polarisation diversity; i.e., it radiates a linearly polarised wave at low frequencies and a circularly polarised wave at high frequencies. In this section we focus our attention on a four-arm spiral antenna and realise another polarisation diversity with all the four arms being fed with the frequency fixed at 3 GHz.

If the antenna periphery is large enough to contain the region of the three-wavelength circumference, the rotational sense of the circular polarisation is uniquely determined by the winding sense of the arms. Since the periphery of the present antenna is 2.4 λ at 3 GHz, namely, the antenna does not contain the three-wavelength circumference, we have the possibility of controlling the sense of a circularly polarised wave by changing the excitation phase.

There are three modes, or phase combinations for exciting the four arms: (1, -j, -1, j), (1, -1, 1, -1), and (1, j, -1, -j). These are called mode 1, mode 2, and mode 3, respectively. In the following analysis we use only two modes, mode 1 and mode 3. The results of the excitation of mode 2 are available in Reference [3].

In the excitation of mode 1, the currents travelling toward the arm ends become in-phase on the region of the one-wavelength circumference, and generate the first mode radiation. Owing to the radiation, the current for the mode 1 decays exponentially from the feed point toward the arm end, as shown in Fig.5.7(a). In contrast, we find that the current distribution by the excitation of mode 3 contains a standing wave over the region from the one-wavelength circumference to the arm ends [see Fig.5.7(b)]. In addition, we find that the standing wave diminishes near the feed point. The behaviour for the mode 3 is similar to that described in Section 5.3.2 and is summarised as follows: (1) the travelling currents without decay pass the region of the one-wavelength circumference without generating an in-phase relation and reach the arm ends without decay; (2) at the arm ends the currents are reflected, and they travel toward the feed points; (3) when the currents pass the region of the one-wavelength circumference, they generate the in-phase relation, or the active region; (4) therefore, circularly polarised radiation generated by these reflected currents has the rotational sense which is opposite to that for the excitation of mode 1; (5) since the reflected currents decay due to the radiation, the standing wave diminishes.

Fig.5.8 shows the radiation pattern and axial ratio when the antenna is excited by the mode 1 and mode 3. Since the active regions for both modes are on the one-wavelength circumference, the radiation patterns are similar to each other. The half-power beamwidth is calculated to be ±40° for the excitation of mode 1 and ±33° for the excitation of mode 3. At $\theta=0°$ (on the Z-axis), the radiation field for each mode has perfect circular polarisation. The angle range in which the radiation is circularly polarised within an axial ratio of

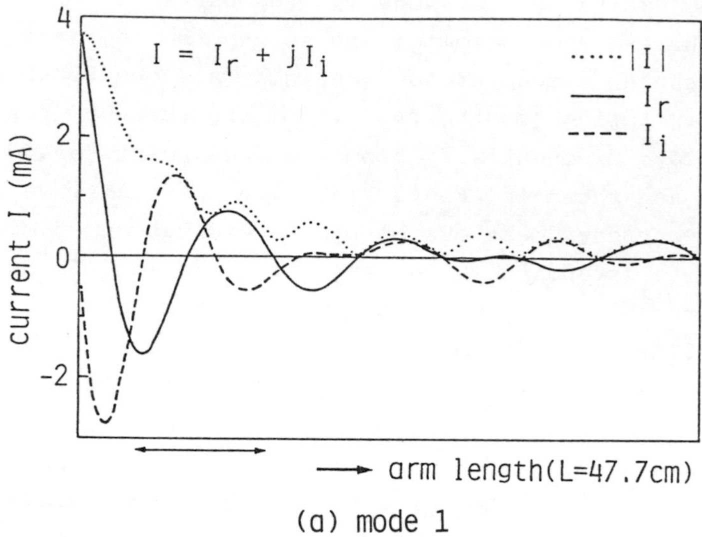

(a) mode 1

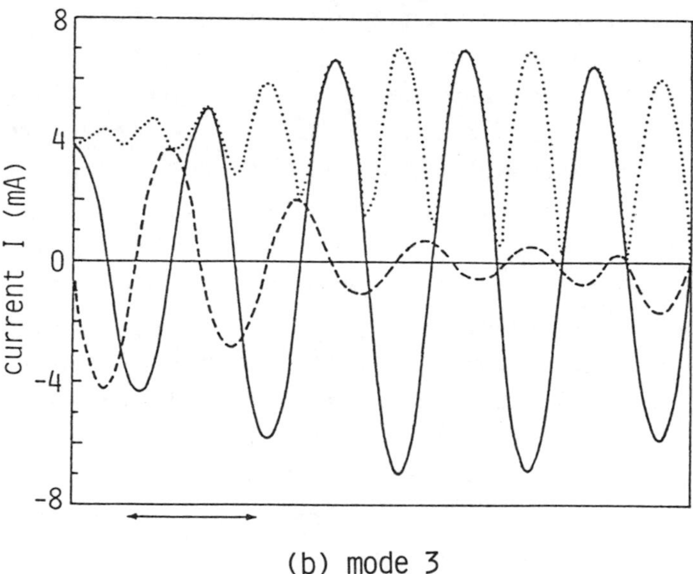

(b) mode 3

Fig.5.7 Current distributions when the four arms are excited.

⟷ the one-wavelength circumference

3 dB is found to be wide: ± 55° for the excitation of mode 1, and ±62° for the excitation of mode 3.

The polarisation diversity described here can be used in a wide range of frequencies from 3 to 3.5 GHz (the antenna periphery 2.4λ∿2.8λ), because the antenna is operated by nonresonant travelling currents.

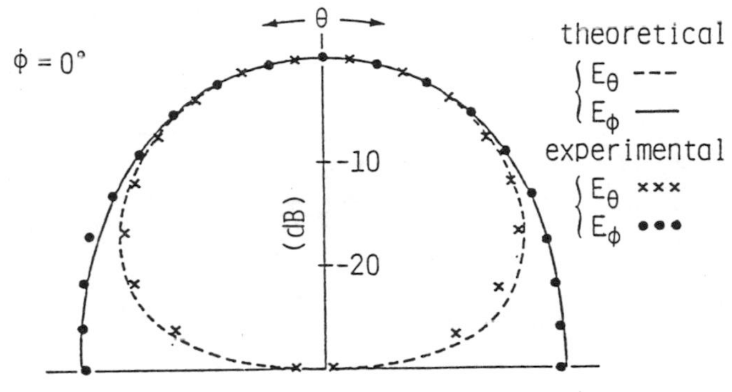

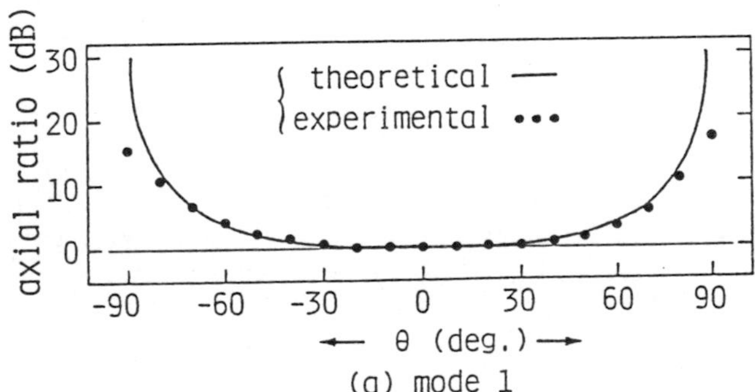

Fig.5.8(a) Radiation pattern and axial ratio when the four arms are excited.

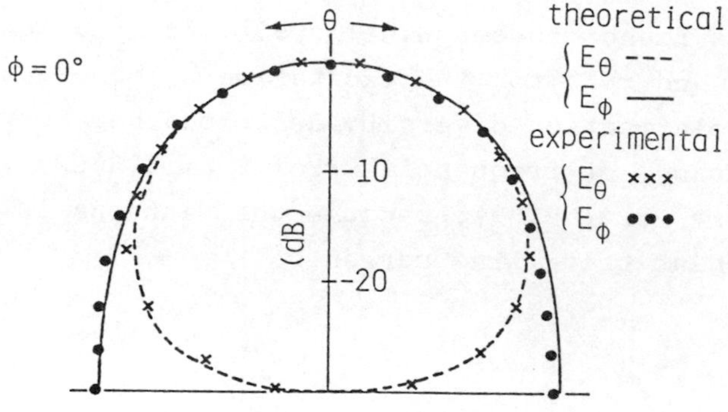

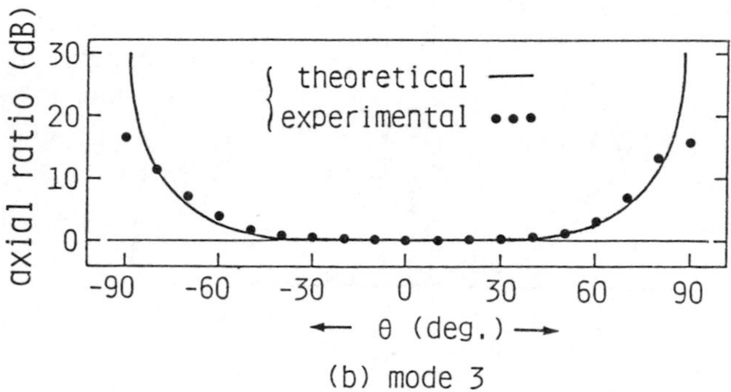

Fig.5.8(b) Radiation pattern and axial ratio when the four arms are excited.

REFERENCES TO CHAPTER 5

[1] Kuo, S.C and Liu, C.C., "Multiple polarization spiral antenna", U.S.Patent 3562756 (Feb. 1971).

[2] Nakano, H., Yamauchi, J. and Sugiyama, Y.,"Parasitic effects and polarization diversity of archimedean spiral antenna", Trans. IECE Japan, J66-B, 11, 1983, pp.1418-1425.

[3] Nakano, H., Yamauchi, J and Hashimoto, S., "Numerical analysis of 4-arm Archimedean spiral antenna", Electronics Letters, Vol.19, 3, 1983, pp.78-80.

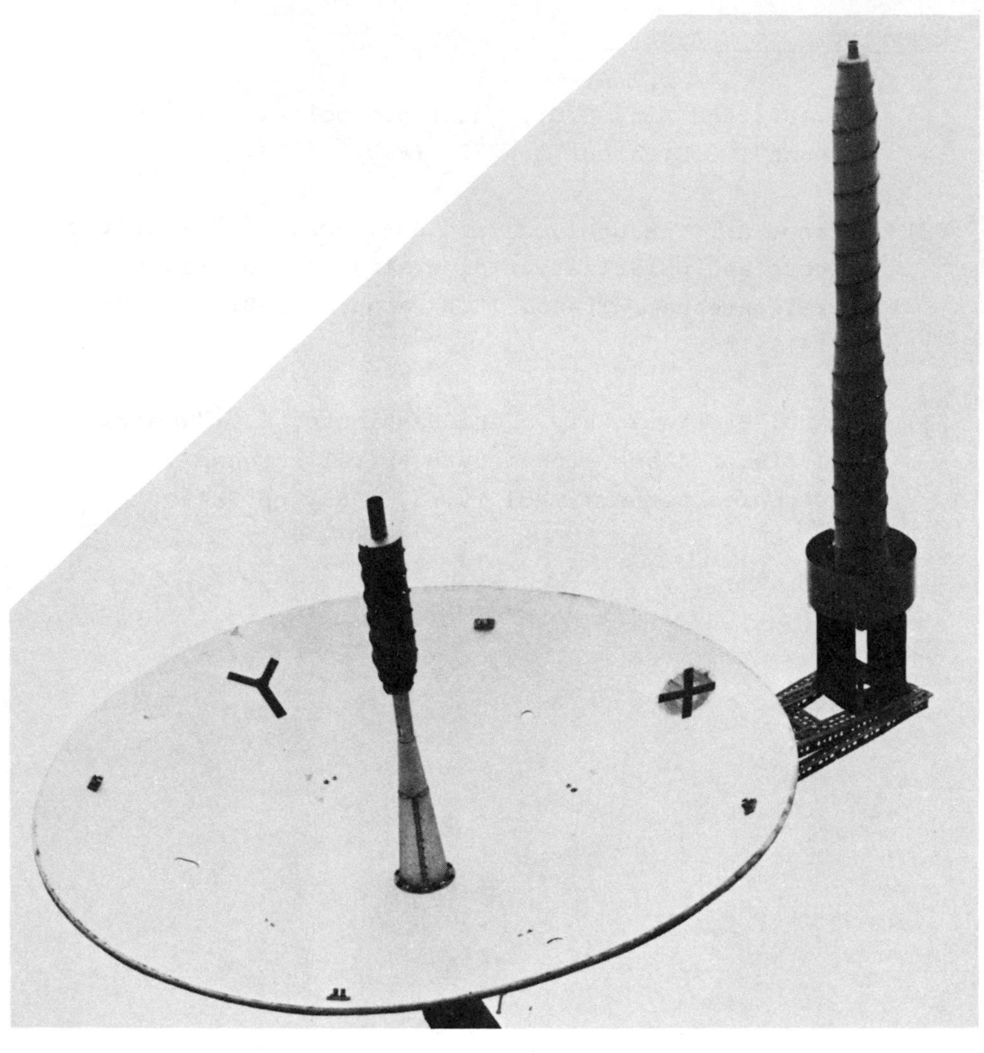

240-400 MHz Antenna system for the fleetsatcom satellites (courtesy of Dr. King, THE AEROSPACE CORPORATION, Ref.[19] on Page 170)

CHAPTER 6
Helical Antenna of Endfire Mode*

6.1. INTRODUCTION

Helical antennas [1][2] have widely been used as simple and practical radiation elements of linear polarisation and circular polarisation. In this chapter we focus our attention on a helical antenna which radiates a circularly polarised wave in the direction of the helical axis. Emphasis is placed on the understanding of the radiation characteristics over a wide frequency range.

In Section 6.3 we discuss the behaviour of the current distribution on a long helical antenna. In the following section we refer to the radiation mechanism.

In Section 6.5 we study an application of a balanced-type helical antenna to the ellipticity measurement of an unknown incident wave. This antenna gives us information about the intensity of a circularly polarised wave of one rotational sense (right-handed or left-handed) relative to the intensity of a circularly polarised wave of the other rotational sense.

Section 6.6 is concerned with the improvement of the axial ratio [3]-[5]. An excellent axial ratio can be realised in practice by tapering the open end section, and the reason for this is clearly given by the theoretically determined current distribution.

Section 6.7 covers the radiation characteristics of

*From papers (p-1),(p-4),(p-5),(p-6),(p-7),(p-11),(p-20),(p-21),(c-3),(c-5),(c-6),(c-8),and (c-14) listed at the end of this monograph.

a 1.5-turn helical antenna. The antenna has an exceptionally good axial ratio despite the short axial length. Therefore, it is worthwhile to use a short helical antenna as an array element. In Section 6.8, arrays for circular polarisation are constructed with 5, 7, and 9 helices, where all the helices are fed from a single waveguide.

6.2. COMPARISON OF TWO TYPES OF BALANCED HELICAL ANTENNAS

We consider the difference of the radiation characteristics between the two types of balanced helical antennas BHA's [6][7]: H.O.(helical arms are wound in the opposite direction) and H.S.(helical arms are wound in the

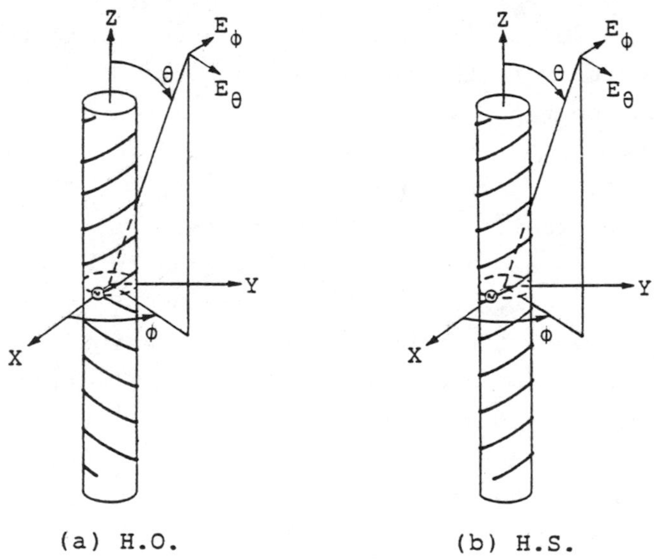

Fig.6.1 Co-ordinate system of balanced helical antennas(BHA's).

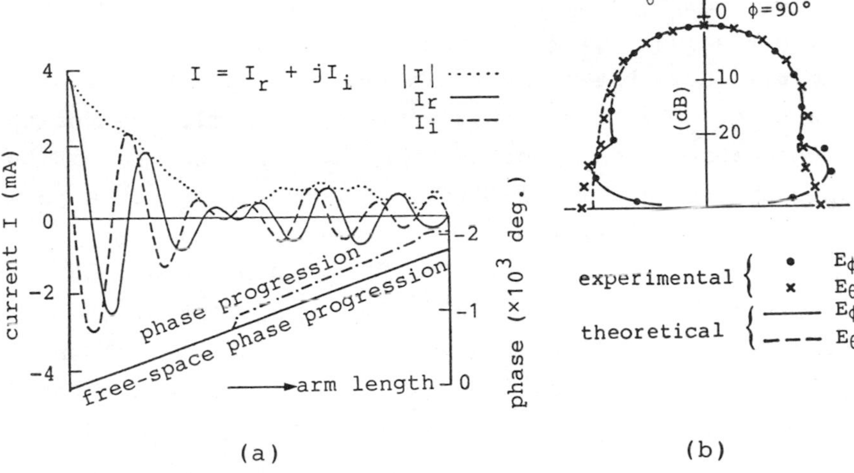

Fig.6.2 Current distribution and radiation pattern of the H.O.

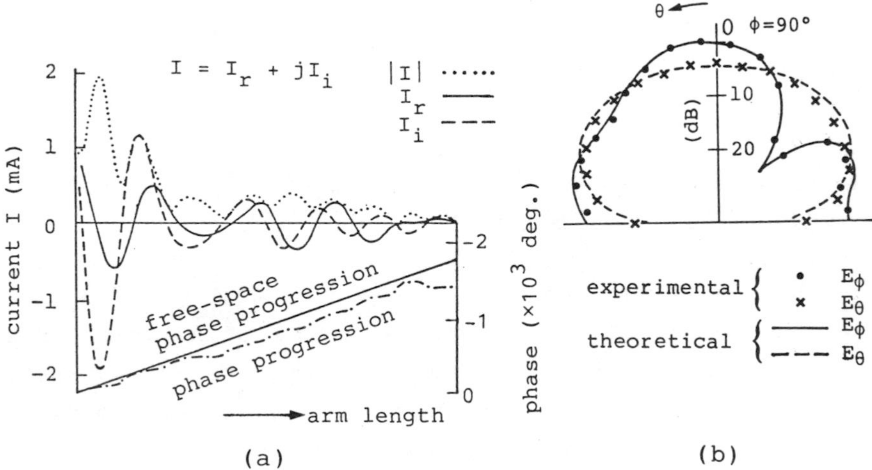

Fig.6.3 Current distribution and radiation pattern of the H.S.

same direction) shown in Fig.6.1. Both helices have the same dimensions: circumference of helical cylinder $\bar{c}$=10 cm, half arm length L=50 cm (the number of turns n≃5), pitch angle α=12.5°, and wire radius ρ=0.5 mm.

Fig.6.2 shows the current distribution and the radiation pattern of the H.O. at 3 GHz. The amplitude of the current attenuates smoothly from the input to the minimum point and reaches a relatively constant value at a point of about 3λ (λ = free-space wavelength) from the input. In this case the axial ratio of the H.O. is 0.7 dB. On the other hand, the amplitude of the current of the H.S. varies greatly near the input region as shown in Fig.6.3(a). This causes the radiation wave to be considerably elliptical. The axial ratio of the H.S. is 9.5 dB at 3 GHz.

When the radiation pattern of the H.O. is compared with that of the H.S., it is obvious that the H.O. has a better symmetrical pattern with respect to the Z-axis. Because of superiority of the H.O. over the H.S. in the radiation characteristics, a BHA of H.O. type is further discussed in the following section.

6.3. RADIATION CHARACTERISTICS OF THE BALANCED HELICAL ANTENNAS RADIATING IN THE AXIAL MODE

Fig.6.4 shows the typical current distribution of a long balanced helix. In this case the number of helical turns is n=22. The calculation indicates that a region S, where the current is travelling with a relatively constant amplitude, expands as the number of helical turns is increased, while a region C, where the current attenuates smoothly into the minimum, is almost independent of the increase of turns [7][8]. The phase velocity in the region S is less than that of light in free space.

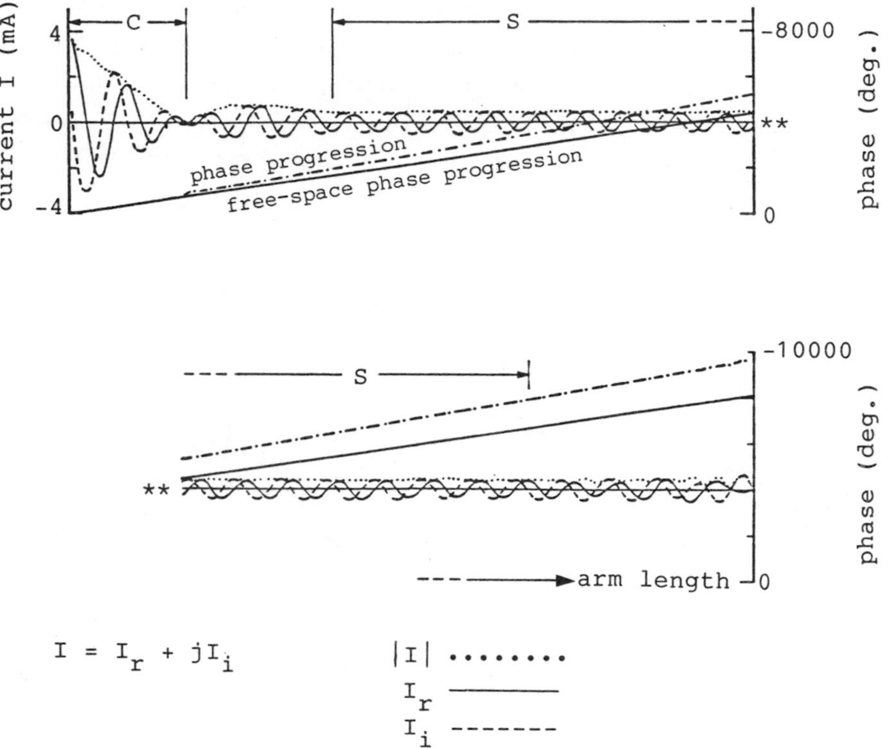

Fig.6.4 Current distribution; n=22, α=12.5°, 3GHz.

Although the change of the pitch angle causes variation in the relative phase velocity p (= phase velocity along the helix/velocity of light in free space), the velocities in the regions C and S are independent of the number of turns n, provided that n is more than 6.

Fig.6.5 illustrates the radiation patterns of helices for different turns. Since the radiation pattern is symmetrical with respect to the XY-plane, only half of the pattern is illustrated. As the number of turns is increased, the radiation beam becomes sharper and the power gain increases. For example, a 16-turn 12.5° helix has a half-power beamwidth of 34° and a power gain of 10.5 dB.

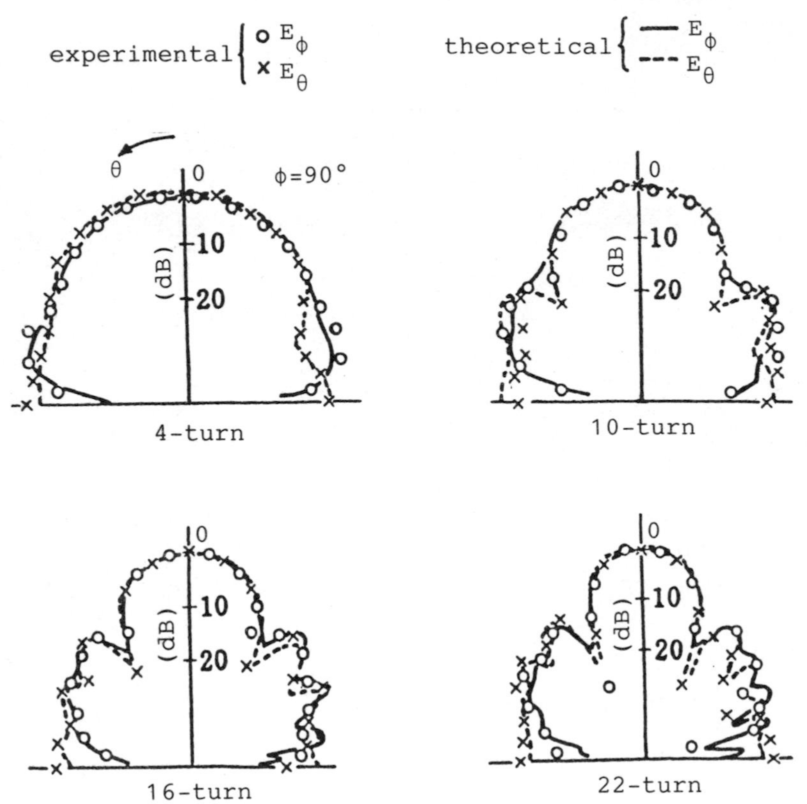

Fig.6.5 Radiation patterns; α=12.5°, $\bar{c}$=10cm, 3GHz.

Fig.6.6 shows the behaviour of the axial ratio with increase in the number of turns. It can be observed that the axial ratio has a cyclical variation rather than decreasing monotonically. The cyclical variation is caused by the reflected currents near the arm ends. The detailed discussion on the mechanism of the cyclical variation is made in Reference [9]. Fig.6.6 also reveals that a 2-turn helix exceptionally radiates a circularly polarised wave in spite of the short axial length. Investigation of a short helix will be thoroughly made in Section 6.7 and its application will be described in Section 6.8.

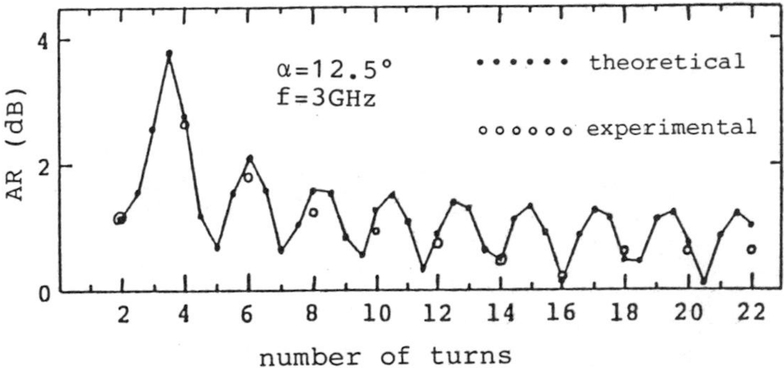

Fig.6.6 Axial ratio with increase in the number of helical turns.

Next we consider the frequency characteristics of a 16-turn 12.5° helix. The variation of the current distribution is shown in Fig.6.7; (a) is at 2.6 GHz ($\bar{c}$=0.867λ) and (b) is at 3.5 GHz ($\bar{c}$=1.167λ). The tendency of the amplitude of the current is towards higher values as the frequency is increased. This tendency is particularly observed in the region S.

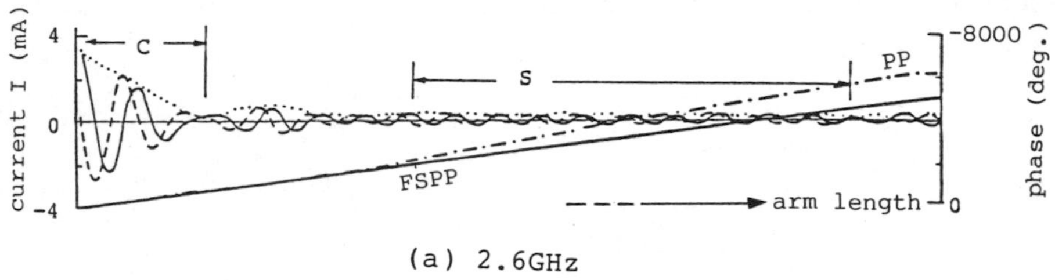

(a) 2.6GHz

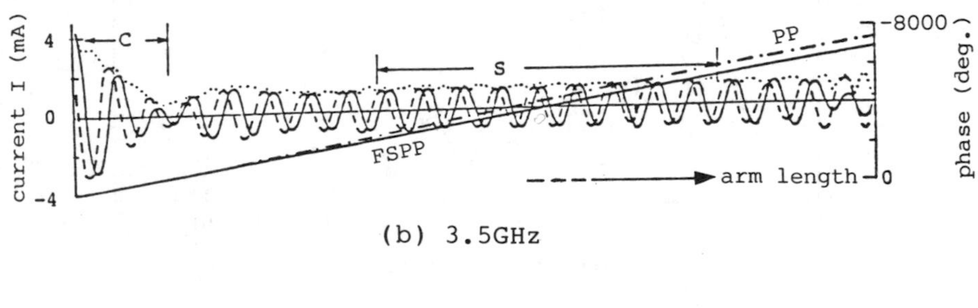

(b) 3.5GHz

$I = I_r + jI_i$ $|I|$ ...... I_r —— I_i - - -

PP: phase progression

FSPP: free-space phase progression

Fig.6.7 Current distributions; n=16, α=12.5°, $\bar{c}$=10cm.

Fig.6.8 shows the relative phase velocities in the regions C and S [7][8][10]. For reference, the phase velocity which satisfies the in-phase condition in the end-fire radiation and that of the Hansen-Woodyard condition are also shown. As a result, it is found that the value of p in the region S changes so that the fields from each turn may add nearly in-phase in the axial direction. This accounts for the persistence of the axial mode of radiation over a relatively wide frequency range.

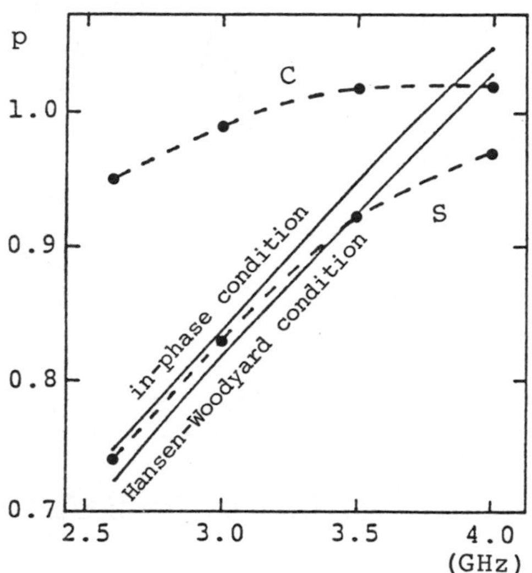

Fig.6.8 Relative phase velocity; n=16, α=12.5°, $\bar{c}$=10cm.

Fig.6.9 shows the radiation patterns at 2.6 GHz and 3.5 GHz. It is seen that many sidelobes appear as the frequency is increased. However, we can find an increase in the power gain due to the fact that the main beam becomes sharper. The power gain of 9.4 dB at 2.6 GHz changes into 11.1 dB at 3.5 GHz.

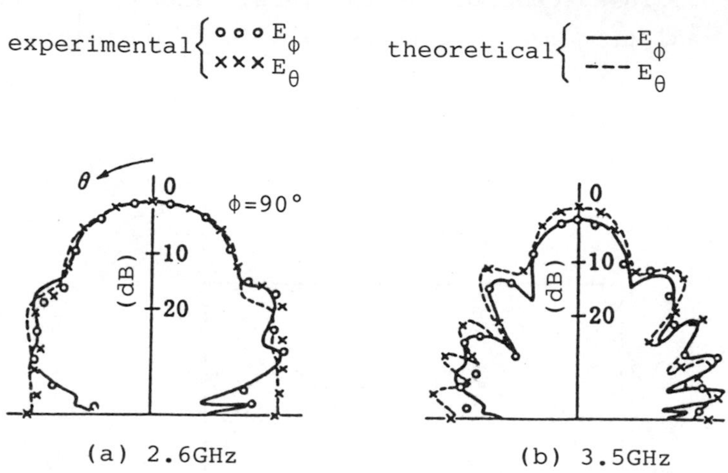

Fig.6.9 Radiation patterns; n=16, α=12.5°, $\bar{c}$=10cm.

6.4. RADIATION MECHANISM OF BALANCED HELICAL ANTENNA [11]

The numerical analysis in the previous section shows that a region S expands as the number of helical turns is increased, while a region C is almost independent of the increase of helical turns. From this result we infer that the region C is a portion where the input power couples into a surface wave, and that the radiation when the helical wire is cut off at the endpoint of the region C is similar to that of a continuously wound balanced helix antenna (CWBH). In this section, in order to confirm this inference, a driven balanced helix with two closely spaced parasitic helices is numerically analysed and compared with the CWBH.

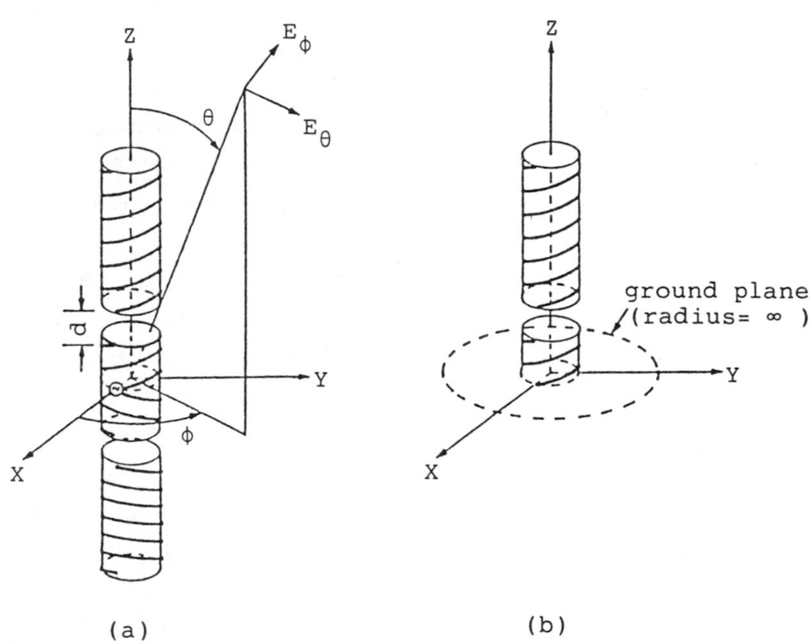

Fig.6.10 Configuration and co-ordinate system.

We cut a helical wire at the endpoint of the region C and obtain an antenna system. The antenna system to be considered here is as follows: a balanced helix of 2 turns is the driven element for the remaining two parasitic helices of 6 turns. Fig.6.10(a) illustrates the co-ordinate system of the model antenna system, in which the parasitic elements lie symmetrically with respect to the XY-plane. Hence the antenna system corresponds to a helix with a parasitic helix mounted on an infinite ground plane as shown in Fig.6.10(b).

An operating frequency of 3 GHz (λ=10cm) is used and the dimensions of the antenna system are chosen as follows: circumference of the helical cylinder $\bar{c}=1\ \lambda$; pitch angle $\alpha=12.5°$; wire radius $\rho=0.005\ \lambda$; spacing between the two elements $d=0.01\lambda$.

Because of the symmetry of arm configuration, the current has equal amplitude and is 180° out of phase with respect to the feed point. Fig.6.11(a) shows the current distribution along the two elements at 3 GHz. It is worth mentioning that, although the helical wire is cut off, the behaviour of the current distribution of this antenna system is similar to that of the 8-turn CWBH shown in Fig.6.11(b). The current travels with a phase velocity nearly equal to the velocity of light along the driven element and travels as a slow wave along the parasitic elements.

The calculated and experimental radiation patterns of the model antenna system are shown in Fig.6.12(a). Excellent agreement between both results is found even in sidelobe levels. The average half-power beamwidth of the E_θ and E_ϕ components is about 55° and the maximum sidelobe level is about -10 dB. Each component shows relatively good symmetry with respect to the Z-axis. As additional information, the radiation pattern of the CWBH calculated by using the current distribution shown in

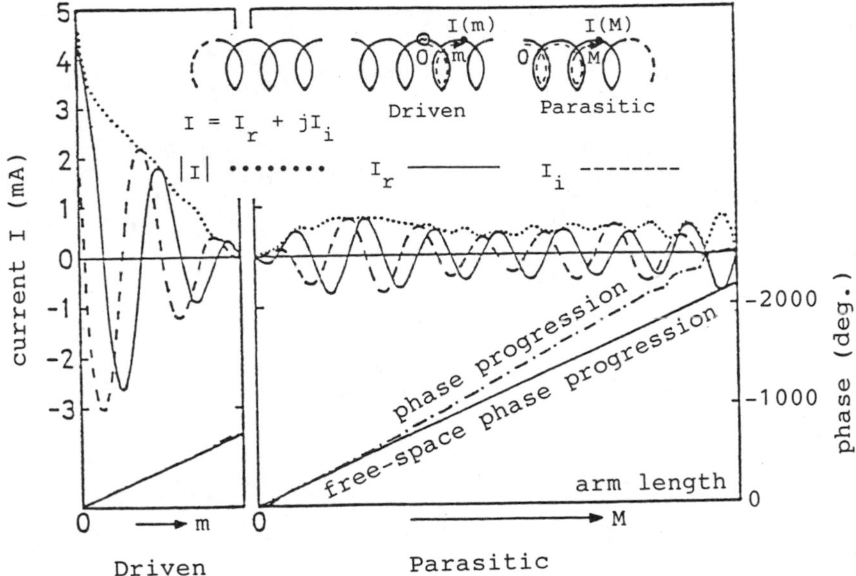

(a) Current distribution of antenna system consisting of driven and parasitic helices.

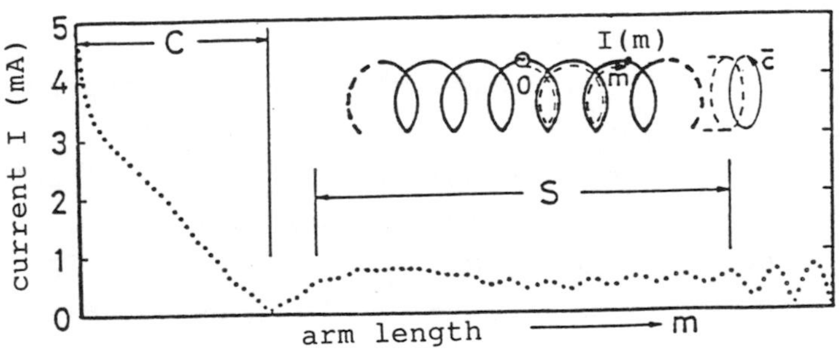

(b) Current amplitude of 8-turn CWBH at 3GHz.

Fig.6.11 Current distributions for two cases.

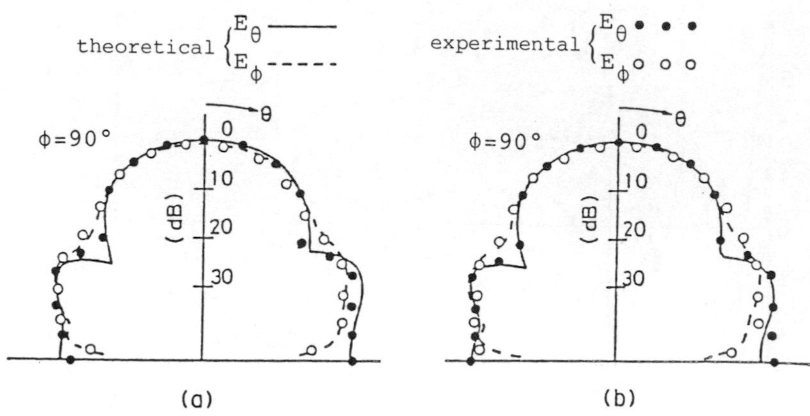

Fig.6.12 Radiation pattern of (a)antenna system and (b)CWBH.

Fig.6.11(b) is also shown in Fig.6.12(b). We can easily find the similarity of radiation patterns between Fig.6.12(a) and Fig.6.12(b).

By using the field components of the E_θ and the E_ϕ in the direction of the helical axis, the axial ratio can be precisely calculated by using eqn.(1.72). The axial ratio of this antenna system is 1.8 dB (measured value = 1.7dB), while that of the CWBH is 1.5 dB (measured value = 1.2dB). With respect to the power gain, the calculated value shows 7.9 dB (measured value = 8.3dB) and it is nearly the same as the power gain of the CWBH, 8.2 dB (measured value = 8.6dB).

In summary, the aforementioned similarities confirm the fact that the first 2 turns of a long CWBH act as an exciter for the remaining turns.

6.5. ELLIPTICITY MEASUREMENT OF AN INCIDENT WAVE [9]

So far, the characteristics of the balanced helical antenna BHA have been described. In contrast with a conventional helix mounted on a ground plane, the BHA can radiate a bidirectional beam. The rotational senses of the beams are opposite to each other in the +Z-axis and -Z-axis directions. Therefore, when the BHA is used as a receiving antenna, right-handed and left-handed circular components of wave, E_R and E_L, can be detected separately but not simultaneously. For the BHA shown in Fig.6.13 the E_R and E_L can be detected when the helical arms A and B are directed toward an incident wave direction, respectively. Then, the ellipticity of wave can be calculated by 20 $\log|(E_R + E_L)/(E_R - E_L)|$ (dB).

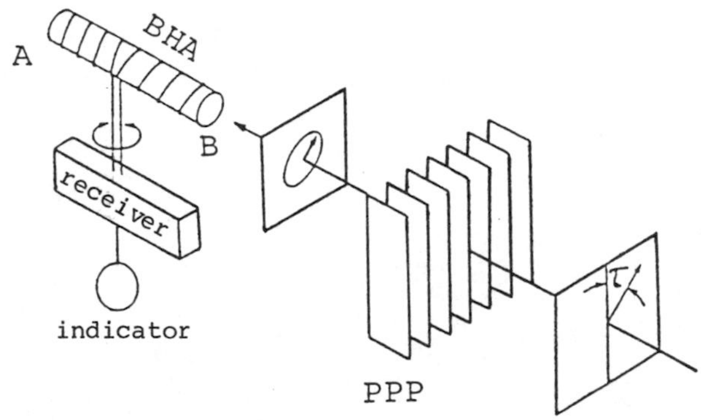

Fig.6.13 Experiment for measuring ellipticity of wave.

The accuracy of this measurement depends on the circularity in the axial ratio of the antenna to be used. Kraus introduced this type of measurement using two long helical antennas: one is wound left-handed and the other is wound right-handed [1]. In this case, the mutual coupling between the two helical antennas should be taken into account with a view to realising an excellent axial ratio. The use of the BHA essentially eliminates this laboriousness.

Fig.6.13 shows an experiment for measuring the ellipticity of wave. A 19-turn BHA of α=13.5° and $\bar{c}$=1 λ is chosen for the measurement because its axial ratio is 0.06 dB which can be regarded as entirely circularly

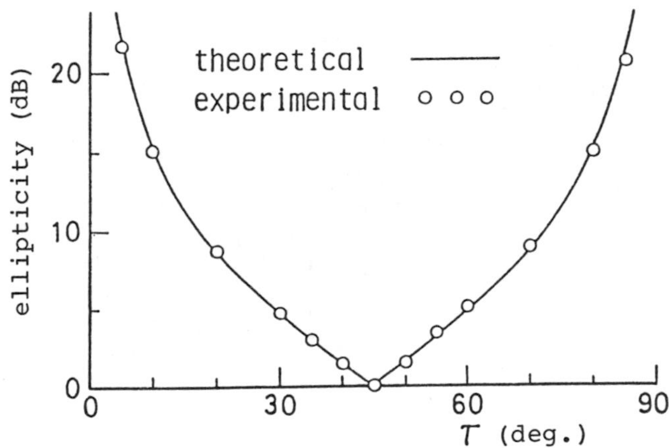

Fig.6.14 Ellipticity ratio of wave.

polarised [9]. A testing wave is generated by a combination of a linearly polarised antenna and a parallel-plane polariser(PPP). A linearly polarised wave coming through the PPP can be changed to a wave of an arbitrary ellipticity by the adjustment of the incident angle τ of the wave. Complete circular polarisation is obtained only when $\tau=45°$.

Fig.6.14 shows an experimental result at 3.0 GHz. It is confirmed that the measured values agree well with a theoretical curve which is obtained on the assumption that the axial ratio of the BHA is 0 dB, i.e., entirely circularly polarised.

6.6. CHARACTERISTICS OF MODIFIED HELICAL ANTENNA [12]

As can be seen from the current distribution shown in Fig.6.4, the last 2-turn is a region where the standing wave is noticeable. Since the deterioration of the axial ratio is mainly due to the reflected current from the open end, it is necessary to suppress the standing wave near the arm end. For this purpose, recent attention is directed to the modifications of the helical arm [3][4][5]. Therefore, this section is devoted to an extensive study of a modified BHA whose open end is tapered. The antenna is designated as TOH.

Fig.6.15 shows the configuration and the co-ordinate system of the TOH which consists of tapered and uniform sections. The tapered section of the TOH is specified by $r=r_0e^{-a\phi_c}$ and the geometric parameters are chosen as follows: generating line of conical $r_0=7.35$ cm, spiral constant $a=0.048$/rad, tapering cone angle $\theta_0=12.5°$ and the number of turns $n_t=2$. Information of θ_0 is available in Reference [13]. The circumference of the helical cylinder in the uniform section is $\bar{c}=10$ cm. In both

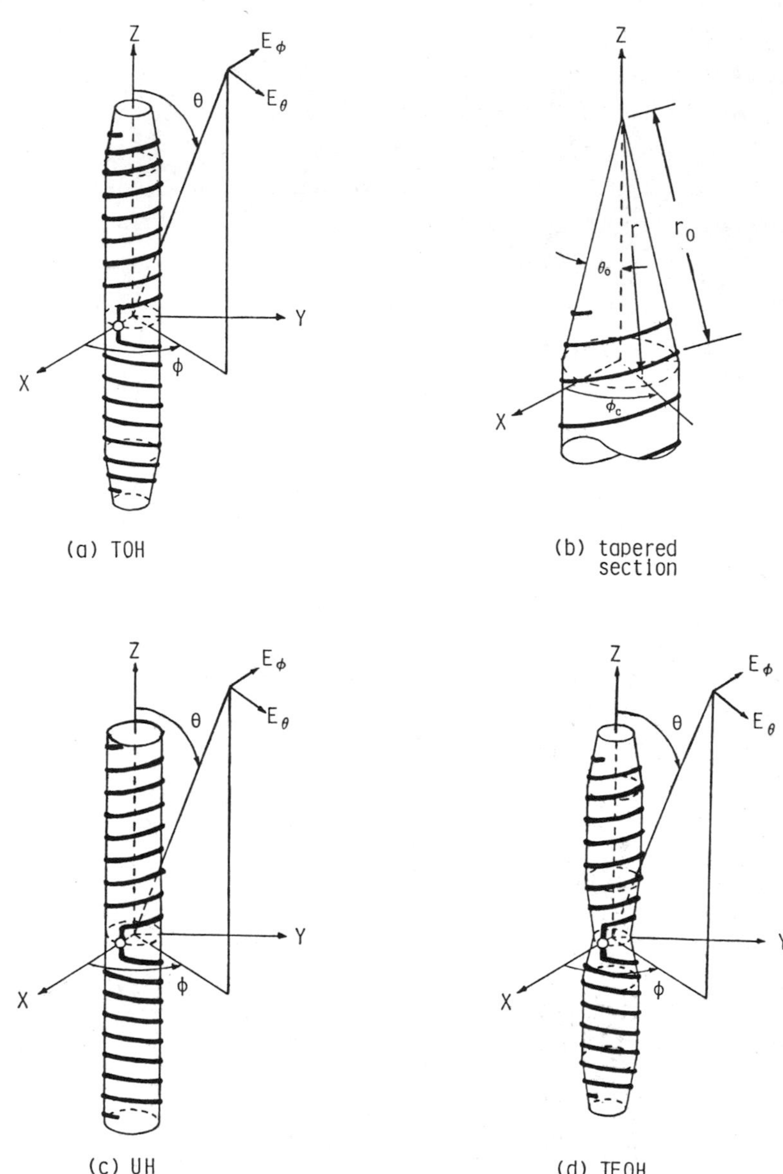

Fig.6.15 Configuration and co-ordinate system of TOH, UH and TFOH.

sections, the pitch angle and the wire radius are taken to be $\alpha=12.5°$ and $\rho=0.5$ mm, respectively. For comparison with the TOH, the characteristics of UH (uniform helix) and TFOH (helix with tapered feed and open end) shown in Figs.6.15(c) and (d) are also investigated under the condition that these helices have the same number of turns.

6.6.1. Characteristics of TOH with increase in the number of helical turns

It is found that the current distribution of the TOH is divided into two distinct regions, as in the case of the UH. As a typical example, the current distribution of the TOH of 8 turns at 3.0 GHz is shown in Fig.6.16. The current of the UH near the open end and that of the TFOH are also shown. It should be noted that the standing wave near the open end of the TOH is greatly suppressed as compared with that of the UH. This fact also holds true for the TFOH. The suppression of the reflected current enables us to predict stability in the input impedance and improvement in the axial ratio.

After tapering the last 2 turns, tapering the first 2 turns has little effect on the main part of the current distribution at 3.0 GHz. The value of the input impedance, however, is influenced by the modification of the helical turns near the feed point.

Fig.6.17 shows the change in the input impedance with increase in the number of turns. As predicted, excellent stability of the input impedance is obtained in the TOH and TFOH. In addition, it is interesting to note that the input impedance for the TFOH is nearly pure resistance.

Fig.6.18 shows a comparison of the axial ratios of three helices. We conclude that the axial ratios of the TOH and

TFOH have excellent stability in comparison with that of the UH. This is due to the fact that tapering the open end greatly contributes to suppression of the reflected current. Since there is essentially no difference in the axial ratios of the TOH and TFOH, it may be said that tapering the feed end has little effect on the axial ratio.

Fig.6.18 also shows the change in the power gain. The increase in the gain results from the fact that the main beam becomes sharper as the number of turns is increased.

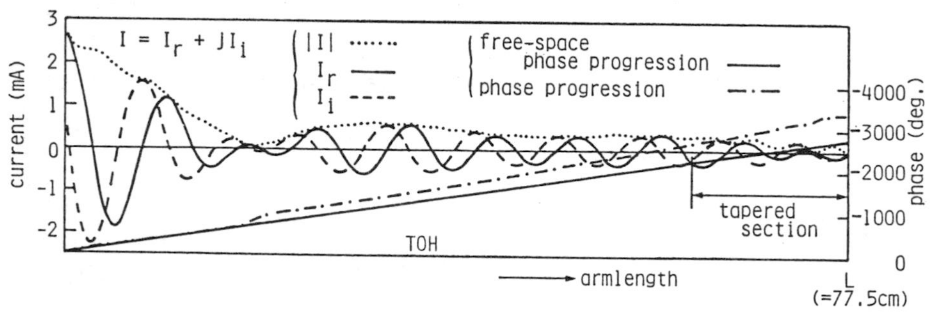

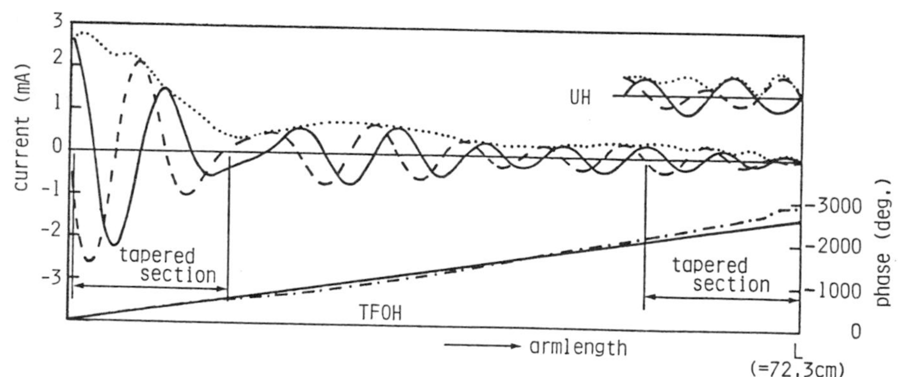

Fig.6.16 Current distributions of TOH, UH and TFOH (8 turns ; 3.0GHz).

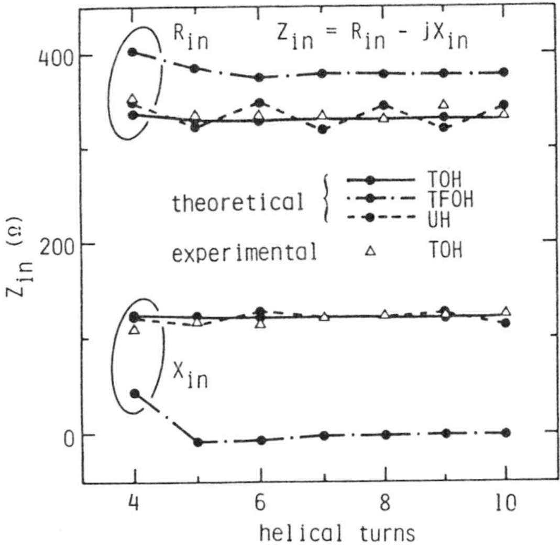

Fig.6.17 Input impedance vs. helical turns (3.0GHz).

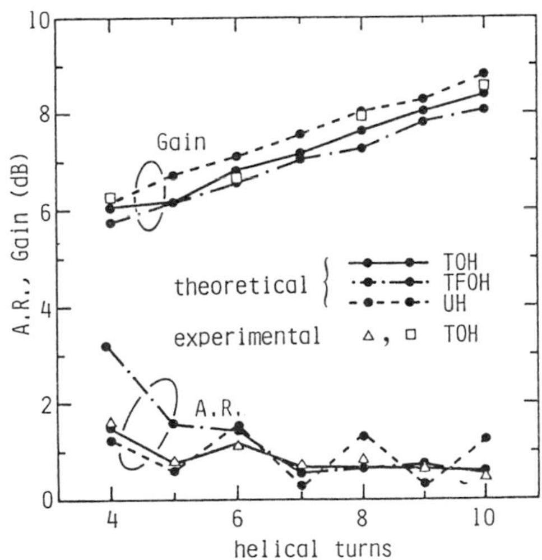

Fig.6.18 Axial ratio and power gain vs. helical turns (3.0GHz).

The overall axial length of the TOH is different from those of the UH and TFOH, and this leads to a slight difference of the power gain, as shown in Fig.6.18. In terms of the axial length, the 2 turns of the tapered section correspond to about 1.5 turns of the uniform section. Generally speaking, when the overall axial length is chosen to be the same, the difference of the gain is negligible.

The typical radiation patterns of the TOH at 3.0 GHz are shown in Fig.6.19. The beam width is a function of the overall axial length, becoming narrower as the number of turns is increased. The patterns of the TFOH and UH of 8 turns are also shown for comparison. Although the form of the sidelobes in the TOH and TFOH is somewhat different from that of the UH, the average half-power beamwidth of each type of helices is nearly the same when each has the same axial length. In the case of the TOH of 8 turns, the beam between half-power points is about 52°. Further calculation shows that the symmetry of the E_θ component with respect to the Z-axis depends on the location and the length of a short straight wire connected to the feed point [see Fig.6.15(a)].

6.6.2. Characteristics of TOH with change in frequency

The current amplitude increases as the frequency is increased. This property is particularly observed in the region S, as shown in Fig.6.20. It should be noted that, due to the tapering effect, the noticeable standing wave observed in the UH does not appear in the current distribution even when the frequency is increased. The current phase velocity in each region is essentially the same as that of the UH.

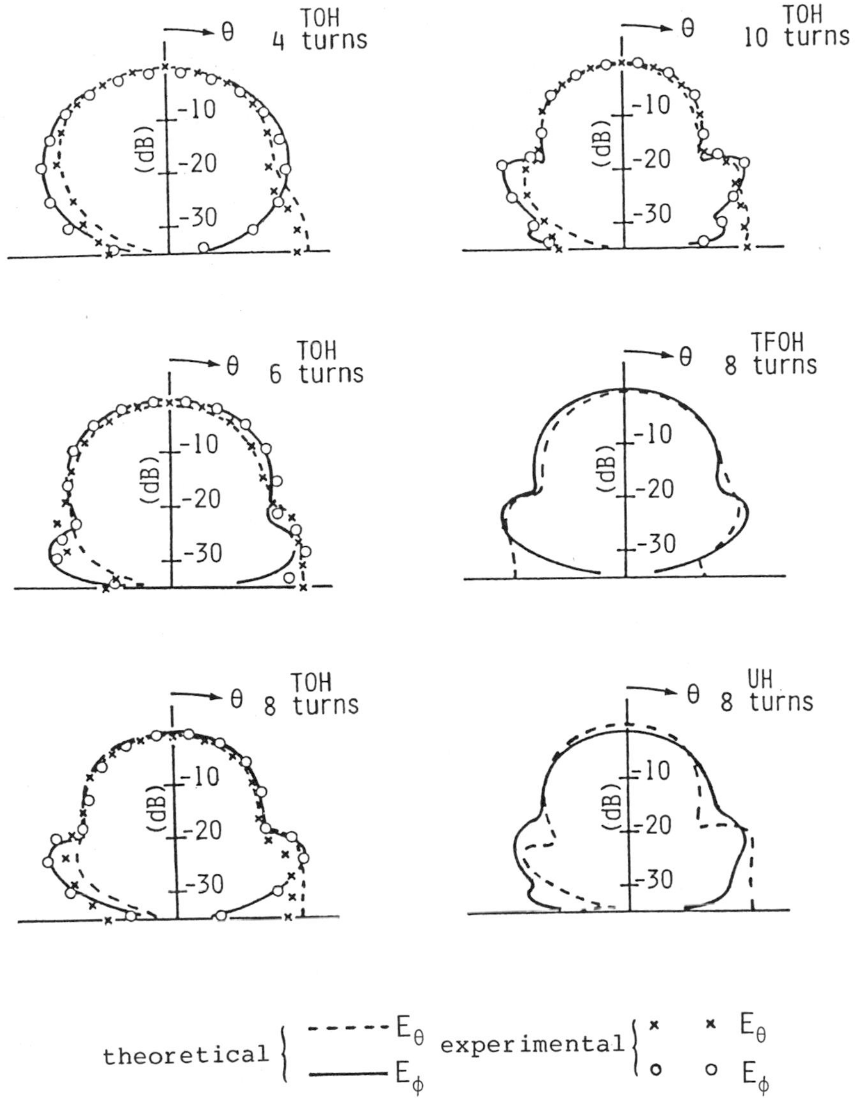

Fig.6.19 Radiation patterns vs. helical turns (3.0GHz ; ϕ=90°).

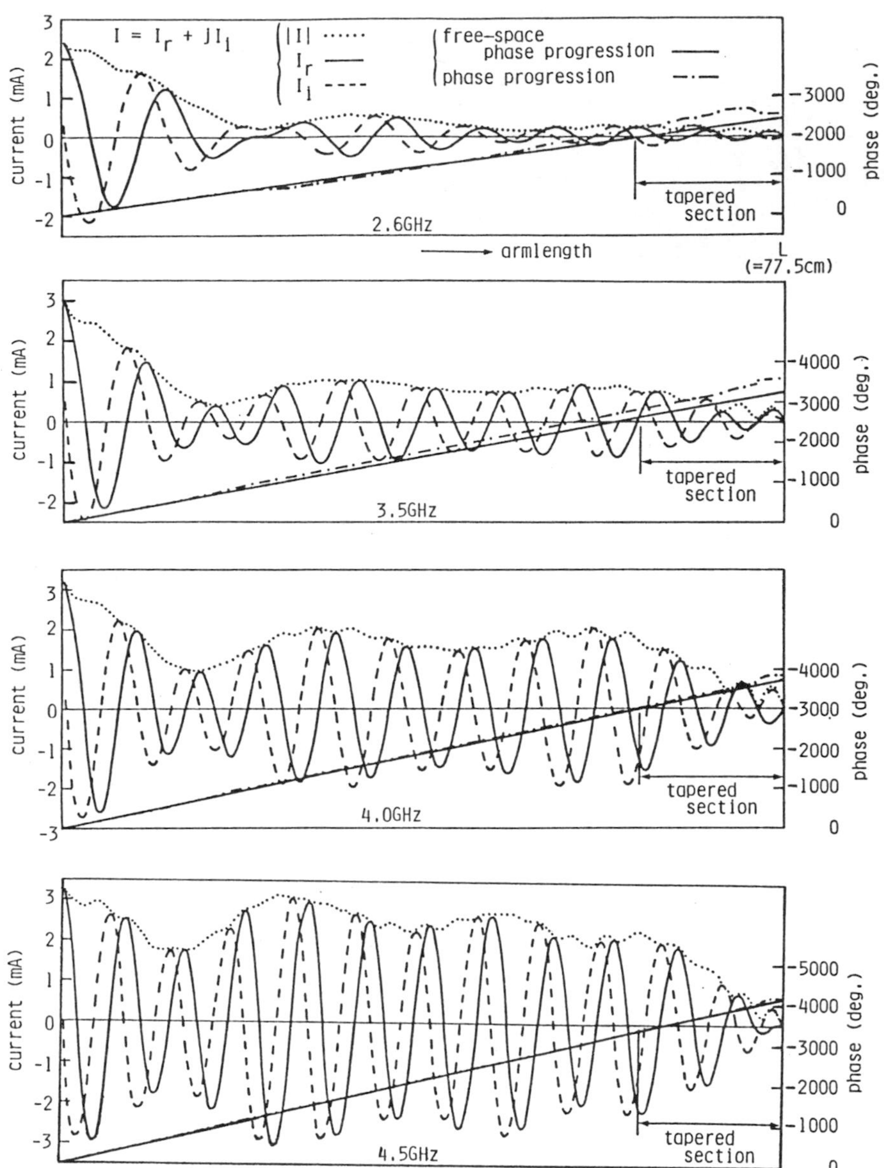

Fig.6.20 Frequency characteristics : current distributions of TOH (8 turns).

The stability of the input impedance is readily achieved, as shown in Fig.6.21. To avoid the overlap of the data plots, only the results for 8 turns are shown. The resistance and reactance of the input impedance of the TOH are, respectively, in the range of 260 ohms to 360 ohms and of -150 ohms to -80 ohms between 2.6 GHz and 4.5 GHz.

Fig.6.22 shows the axial ratio characteristic when the frequency is changed. The axial ratio of the TOH of 8 turns is within 1.5 dB over a range of 2.6 GHz to 4.5 GHz. It can be said that the axial ratio is improved

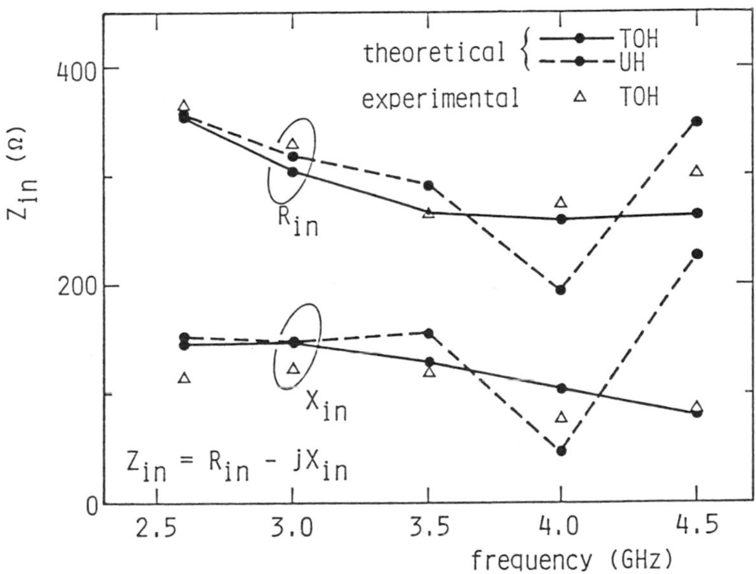

Fig.6.21 Frequency characteristics : impedance (8 turns).

especially at higher frequencies, as compared with that of the UH.

The power gain characteristic is shown in Fig.6.23. The frequency at which the peak gain is obtained is not necessarily the same in the TOH of different turns. For example, the gains of the TOH of 4 turns and 6 turns continue to increase up to a frequency of 4.0 GHz. In contrast, the gains of the TOH of 8 turns and 10 turns at 4.0 GHz are lower than those at 3.5 GHz. This degradation of gain results from the split in the radiation pattern as shown in Figs.6.24(c) and (d).

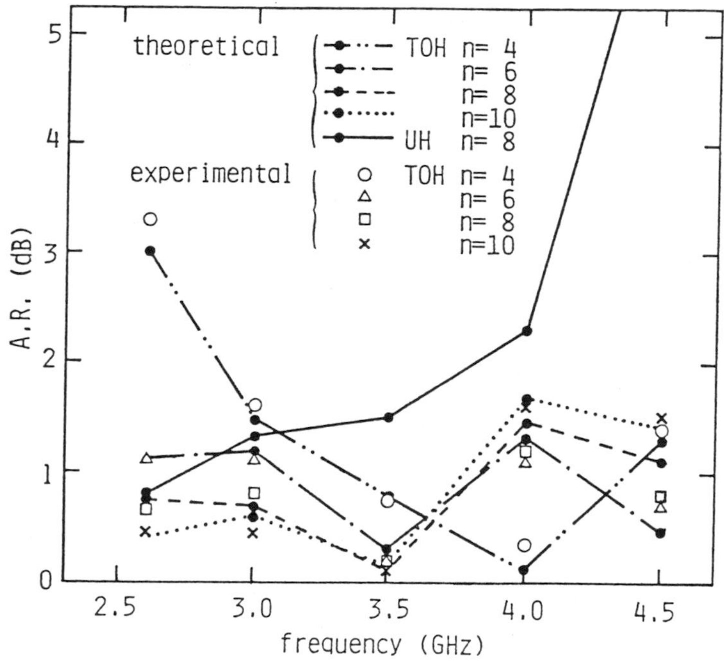

Fig.6.22 Frequency characteristics : axial ratio.

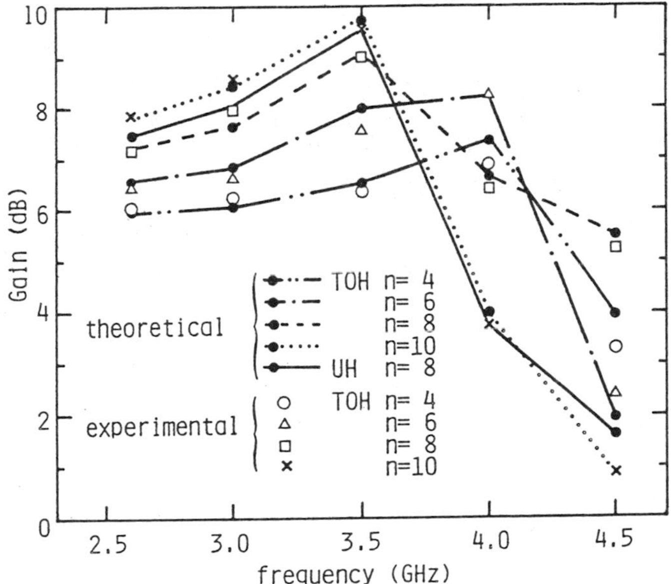

Fig.6.23 Frequency characteristics : power gain.

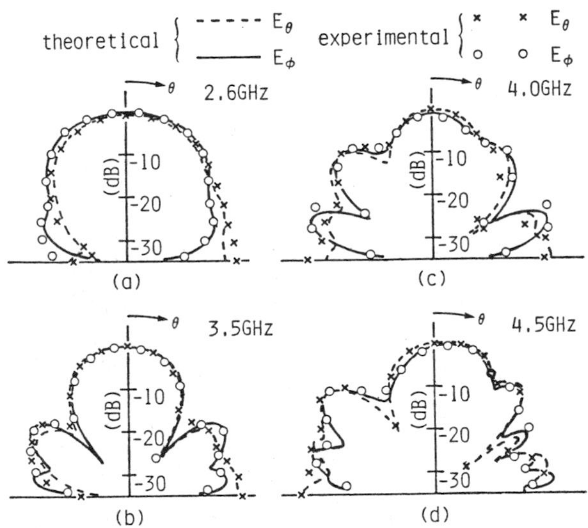

Fig.6.24 Frequency characteristics : radiation patterns of TOH (8 turns ; ϕ=90°).

6.7. CHARACTERISTICS OF SHORT HELICAL ANTENNA [14]

Although the axial ratio tends to deteriorate as the number of turns is decreased, a 2-turn helix fed at its periphery radiates a circularly polarised wave exceptionally well in spite of the short axial length [see Fig.6.6].

The purpose of this section is to investigate the characteristics of a short helix in detail. The feed point of the helix in this section is located on the helical axis so that the helix may rotate about its axis. This facilitates the adjustment of the radiation phase as will be mentioned in Section 6.8.

6.7.1. Radiation characteristics

Fig.6.25 shows the configuration and co-ordinate system. The two antennas of (a) and (b) are electrically the same. The circumference of the helical cylinder $\bar{c}$ and the pitch angle α are chosen to be, respectively, $\bar{c}=1\ \lambda$ and $\alpha=12.5°$, which are common in a conventional axial-mode helical antenna. The wire radius ρ is taken to be $0.005\ \lambda$. The feed point is located on the helical axis, and a short wire is inserted between the feed point and the beginning of the helix proper. This allows the rotation of the helix about the helical axis, with feed point being fixed.

The current distribution of a 1.5-turn helix is shown in Fig.6.26. For reference, the current amplitude of a 1-turn helix is also plotted. It is found that a standing wave distribution for the 1-turn helix disappears when the number of helical turns is 1.5. The current decays smoothly along the arm, travelling with a velocity nearly equal to that of light in free space. It is noted that the decaying current region corresponds to a portion where the

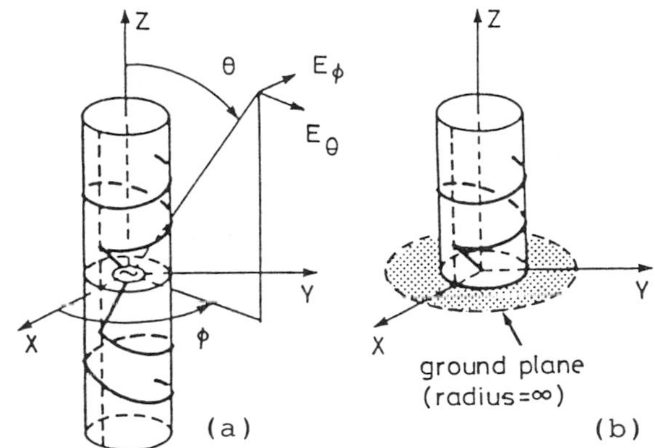

Fig.6.25 Configuration and co-ordinate system.

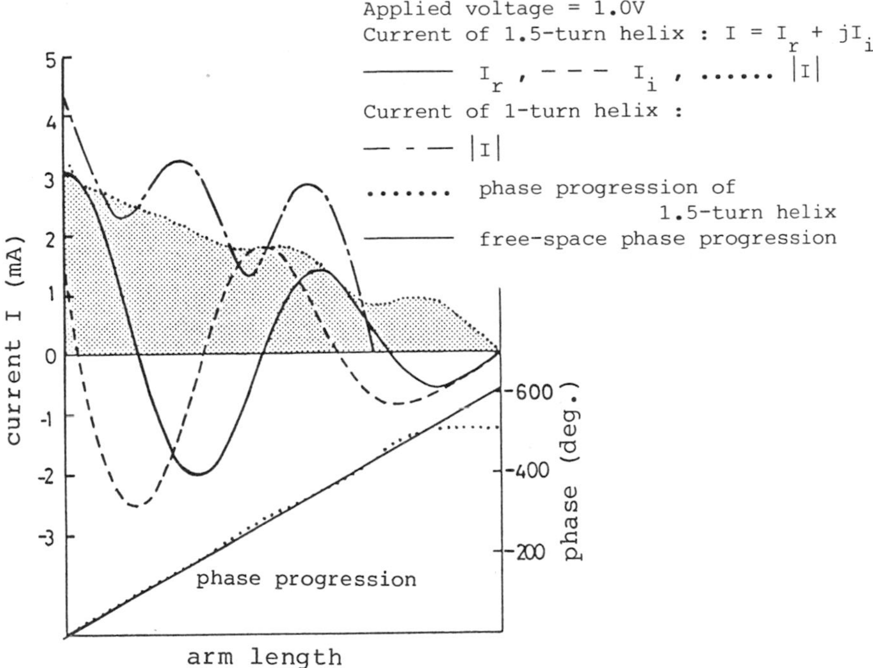

Fig.6.26 Current distributions of 1.5- and 1-turn helices.

input power couples into a surface wave in a long helix. The operation of the short helix characterised by a leaky-wave nature should be distinguished from that of the long helix characterised by a surface-wave nature.

Fig.6.27 shows the calculated and experimental radiation patterns. The radiation beam is found to be broad, and the average half-power beamwidth of the E_θ and E_ϕ components is about 86°. This leads to a low gain of 4.6 dB (measured value=4.9 dB) in the Z-axis direction. In the same direction, the axial ratio is calculated to be 2.6 dB (measured value=2.6 dB).

The fundamental condition required for the radiation of circular polarisation is to distribute a decaying current along a short helix. When we fail to do so, we cannot obtain a good axial ratio. Indeed, the axial ratio of 1-turn helix is elliptically polarised with an axial ratio of 8 dB.

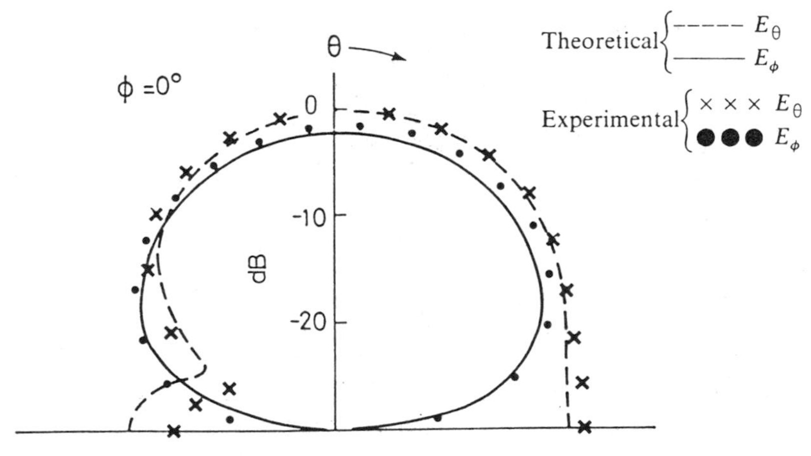

Fig.6.27 Radiation pattern of 1.5-turn helix.

6.7.2. Mutual coupling between short helical antennas

We consider the case where a pair of identical 1.5-turn helices wound in the same sense are arrayed as shown in the inset of Fig.6.28. The interaction between the helices is expressed by the ratio of input currents $|I_1/I_2|$ when helix 1 is excited and helix 2 is connected

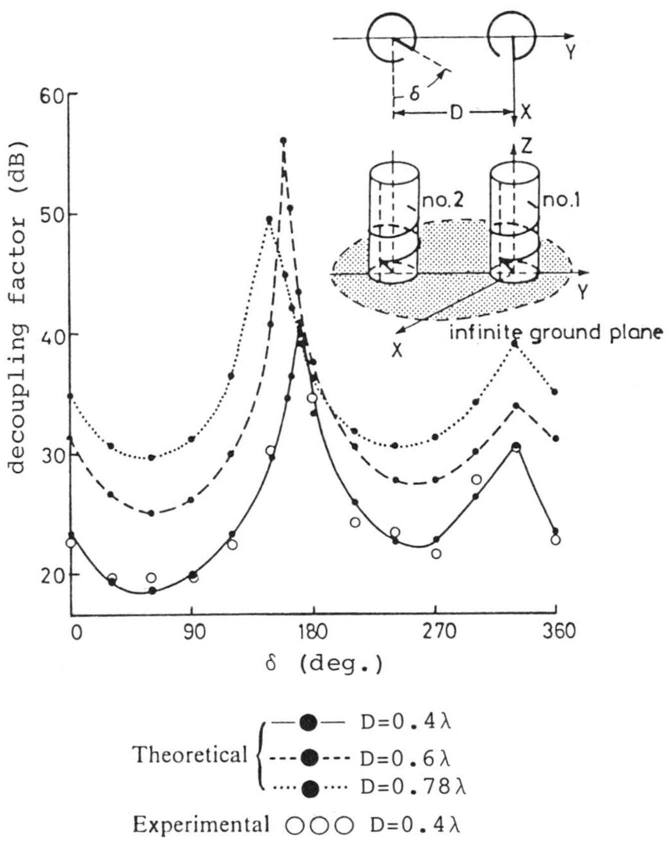

Fig.6.28 Decoupling factor between a pair of 1.5-turn helices.

to a conjugate matched load impedance, i.e.,

$$\left| \frac{I_1}{I_2} \right| = \frac{2 \operatorname{Re} (Z_s)}{| Z_m |}$$

where Z_s is a self impedance of helix 2 and Z_m is the mutual impedance between the helices. The ratio is called a decoupling factor (DCF) [15].

Fig.6.28 shows how the value of DCF varies with the rotation angle δ and the separation distance D. Experimental results for D=0.4 λ are obtained using an open-circuit/short-circuit method. Excellent agreement is seen to exist between the calculated and experimental results. It is found that the value of DCF becomes maximum in the vicinity of δ=180° when the separation distance D is 0.4 λ. Fig.6.28 also indicates that the change in DCF is similar for other separation distances. It should be noted that the rotation of the helix, as well as the increase in the separation distance, is a useful means of reducing the mutual coupling between the helices.

6.8. SHORT HELICAL ANTENNA ARRAY FED FROM A WAVEGUIDE [16]

In this section we study an array of helical antennas fed from a waveguide as shown in Fig.6.29. The array element has a helical section and a linear section. The number of helical turns is taken to be 1.5 turns on the basis of the results of the previous section. The helical section is surrounded with a cavity in order to reduce the mutual couplings among the array elements [17][18]. The linear section is inserted into the waveguide through a small hole and excited by TE_{10} mode in a rectangular waveguide. A portion of the power in the waveguide is transmitted in sequence into the helical sections and is radiated out into free space; the remaining power travels toward the end of the waveguide where it is absorbed by a dummy load. If the array element were to be conventionally fed from a coaxial line, then the number of coaxial lines required by the array would be equal to the number of elements. In addition, the same number of phase shifters and attenuators are required. The use of a waveguide eliminates the disadvantage of the complicated feeding structure.

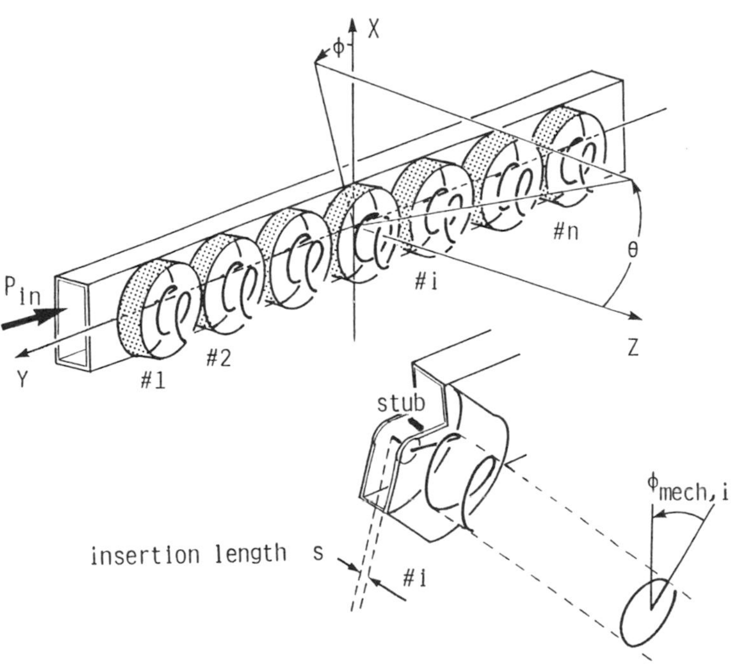

Fig.6.29 Configuration of a helical antenna array fed from a waveguide.

The power efficiency in the array system using the waveguide is defined as $\eta = P_{rad}/P_{in}$, where P_{rad} is the total radiation power from the array elements and P_{in} is the input power to the waveguide. In the case where the input power P_{in} and the power efficiency η are taken to be constant, the radiation power per array element is increased as the number of the array elements is decreased. Consequently, it becomes necessary for each array element to have a longer insertion length. This causes difficulty in designing the array, due to larger reflected waves in the waveguide. To eliminate the difficulty, tuning stubs are introduced near the array elements. This leads to a smooth flow of the power in the waveguide and simplifies the array design.

6.8.1. Design procedure

The design is carried out using transmission line theory. Fig. 6.30 shows the equivalent circuit of the array consisting of n helical antennas, where $Y_{H,i}(=G_{H,i} + jB_{H,i})$ is the admittance of the i-th array element, and $Y_{p,i}(=jB_{p,i})$ is the admittance of the i-th tuning stub. For convenience, a set of $Y_{H,i}$ and $Y_{p,i}$ is called an i-th cell. Each cell is adjusted so that the input power flows smoothly toward the end of the waveguide. In other words, the input impedance of each cell is matched to the characteristic impedance of the waveguide, $Z_0(=1/Y_0)$. The cell in this situation is called a "matched cell".

Coupling phase ϕ_i

We assume that the mutual couplings among the array elements on the waveguide are negligible, and first determine a phase difference θ_i between the terminal voltage of the i-th tuning stub, V_i, and the terminal voltage of the i-th array element, $V_{in,i}$. V_i is given by

$$V_i = (\cos\beta\ell_i + jZ_0Y_{H,i}\sin\beta\ell_i)V_{in,i} - jZ_0I_{in,i}\sin\beta\ell_i \quad (6.1)$$

where $\beta = 2\pi/\lambda_g$ (λ_g is a wavelength in the waveguide) is the phase constant in the waveguide, and ℓ_i is the distance between the i-th array element and the i-th tuning stub. $I_{in,i}$ in the matched cell is

$$I_{in,i} = \frac{V_{in,i}}{Z_0} \ . \quad (6.2)$$

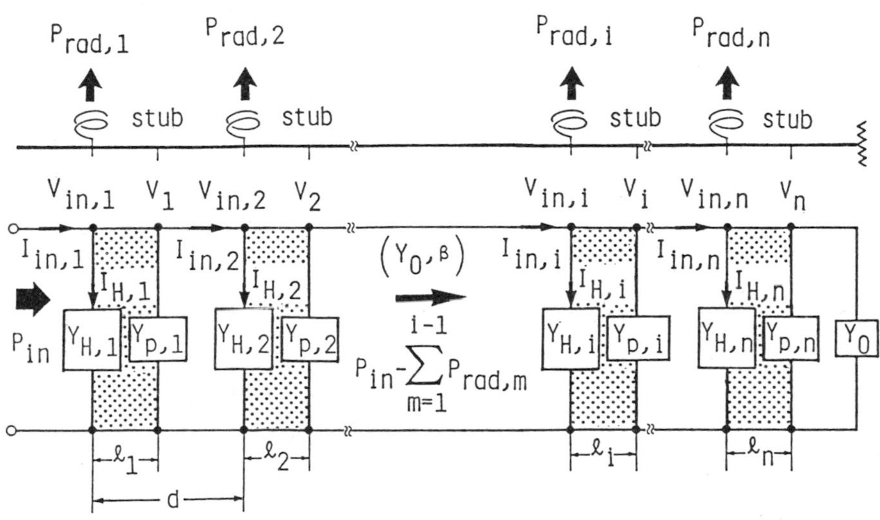

Fig.6.30 Equivalent circuit of an array consisting of n helical antennas.

Substituting eqn.(6.2) into eqn.(6.1), we have

$$\frac{V_i}{V_{in,i}} = [(\cos\beta\ell_i - Z_0 B_{H,i}\sin\beta\ell_i)^2 + (Z_0 G_{H,i} - 1)^2 \sin^2\beta\ell_i]^{1/2}\, e^{j\theta_i} \quad , \tag{6.3}$$

where the phase difference θ_i is

$$\theta_i = \tan^{-1}\frac{(Z_0\, G_{H,i} - 1)\tan\beta\ell_i}{1 - Z_0\, B_{H,i}\tan\beta\ell_i} \,. \tag{6.4}$$

Secondly, we determine the distance between the array element and the tuning stub, ℓ_i. The matching condition of the i-th cell is expressed by

$$Y_0 = G_{H,i} + jB_{H,i} + Y_0\,\frac{Z_0(\,jB_{p,i} + Y_0) + j\tan\beta\ell_i}{1 + jZ_0(jB_{p,i} + Y_0)\tan\beta\ell_i} \,. \tag{6.5}$$

From the identities of the real parts and the imaginary parts in the right and left sides in eqn.(6.5),

$$\ell_i = \frac{1}{\beta}\tan^{-1}\frac{G_{H,i}}{B_{H,i} - \sqrt{(G_{H,i}{}^2 + B^2{}_{H,i})(1 - Z_0 G_{H,i})}} \,. \tag{6.6}$$

Thirdly, we determine the phase of the current flowing to the i-th array element, ϕ_i. The ϕ_i is designated as a "coupling phase".

The current flowing to the i-th array element, $I_{H,i}$, is given by

$$I_{H,i} = (G_{H,i} + jB_{H,i})V_{in,i} \,. \tag{6.7}$$

Equation (6.7) is transformed into eqn.(6.8), using $V_{in,1}$ as a reference voltage:

$$I_{H,i} = \sqrt{(G_{H,i}^2 + B_{H,i}^2)} \; e^{jh_i} \left| \frac{V_{in,i}}{V_{in,1}} \right| e^{j\psi_i} | V_{in,1} | . \quad (6.8)$$

where

$$h_i = \tan^{-1} \frac{B_{H,i}}{G_{H,i}} , \quad (6.9)$$

and ψ_i is a phase difference between $V_{in,i}$ and $V_{in,1}$. $V_{in,i}/V_{in,1}$ in eqn.(6.8) is expressed as

$$\frac{V_{in,i}}{V_{in,1}} = \left(\frac{V_1}{V_{in,1}} \frac{V_{in,2}}{V_1} \right) \left(\frac{V_2}{V_{in,2}} \frac{V_{in,3}}{V_2} \right) \cdots\cdot$$

$$\cdot \left(\frac{V_{i-1}}{V_{in,i-1}} \frac{V_{in,i}}{V_{i-1}} \right) . \quad (6.10)$$

We recall that the waveguide is assumed to be lossless, and that each cell is in the condition of the matched cell. Then,

$$\frac{V_{in,i}}{V_{in,1}} = \frac{V_1}{V_{in,1}} e^{-j\beta(d-\ell_1)} \frac{V_2}{V_{in,2}} e^{-j\beta(d-\ell_2)} \cdots\cdots$$

$$\cdot \frac{V_{i-1}}{V_{in,i-1}} e^{-j\beta(d-\ell_{i-1})} , \quad (6.11)$$

where d is the distance between the array elements. Equation (6.8) is expressed using eqn.(6.11) as follows:

$$I_{H,i} = \sqrt{G_{H,i}^2 + B_{H,i}^2} \left| \frac{V_1}{V_{in,1}} \right| \left| \frac{V_2}{V_{in,2}} \right| \cdots\cdots$$

$$\cdot \left| \frac{V_{i-1}}{V_{in,i-1}} \right| \, | V_{in,1} | \, e^{j\phi_i} , \qquad (6.12)$$

where the coupling phase is

$$\phi_i = \psi_i + h_i = \theta_1 - \beta(d-\ell_1) + \theta_2 - \beta(d-\ell_2)$$

$$+ \ldots + \theta_{i-1} - \beta(d-\ell_{i-1}) + h_i . \qquad (6.13)$$

Finally, substituting eqns.(6.4), (6.6) and (6.9) into eqn.(6.13), we have

$$\phi_i = \sum_{m=1}^{i-1} \Delta_m - (i-1)\beta d + \tan^{-1} \frac{B_{H,i}}{G_{H,i}} , \qquad (6.14)$$

where

$$\Delta_m = \tan^{-1} [-G_{H,m}^2 Z_0 / \{ -B_{H,m}(2-G_{H,m}Z_0)$$

$$+ 2\sqrt{(G_{H,m}^2 + B_{H,m}^2)(1 - G_{H,m}Z_0)} \}] . \qquad (6.15)$$

Coupling factor γ_i

The coupling factor in the i-th array element, γ_i, is defined as the ratio of $P_{rad,i}/P_{in,i}$, where $P_{rad,i}$ is the radiation power from the i-th array element and $P_{in,i}$ is the input power to the i-th array element. In the condition of the matched cell

$$\gamma_i = \frac{P_{rad,i}}{P_{in,i}} = \frac{G_{H,i}|V_{in,i}|^2}{Y_0|V_{in,i}|^2} = G^*_{H,i} \quad , \tag{6.16}$$

where the asterisk is the notation of normalisation to the characteristic admittance of the waveguide, $Y_0(=1/Z_0)$. Since $P_{in,1}$ is equal to P_{in} in the condition of the matched cell, the coupling factor in the first element γ_1 is expressed using the power efficiency η:

$$\gamma_1 = \frac{P_{rad,1}}{P_{in,1}} = \frac{P_{rad}}{P_{in}} \frac{P_{rad,1}}{P_{rad}} = \eta \frac{P_{rad,1}}{\sum_{m=1}^{n} P_{rad,m}} \,. \tag{6.17}$$

The coupling factor in the second element is given by

$$\gamma_2 = \frac{P_{rad,2}}{P_{in,2}} = \frac{\gamma_1 \,(P_{rad,2}/P_{rad,1})}{1 - \gamma_1} \,. \tag{6.18}$$

Generally, the coupling factor in the i-th element is

$$\gamma_i = \frac{\gamma_1(P_{rad,i}/P_{rad,1})}{1 - \gamma_1 \sum_{m=1}^{i-1} (P_{rad,m}/P_{rad,1})} = G^*_{H,i} \,. \tag{6.19}$$

Determination of design parameters

The array of helical antennas is realised by using eqn.(6.14) and eqn.(6.19). Hence, fundamental information on the admittance of the helical antenna in the equivalent circuit presentation, $G^*_H + jB^*_H(=G_H/Y_0 + jB_H/Y_0)$, is required in advance. The admittance of the helical antenna depends on the insertion length of linear section into the waveguide [see Fig.6.29]. The admittance is experimentally obtained by a conventional standing wave method. An example of the admittance graph will be shown in Fig.6.31. Once the relation between the admittance and the insertion length is known, the insertion length which satisfies a specified value of conductance G_H^* can be determined, giving a value of susceptance B_H^*. After preparing the admittance graph, we calculate the design parameters as follows.

First, the power efficiency η is chosen to be a specific value in eqn.(6.17). Secondly, we give the aperture distribution or the radiation power distribution at the aperture, $P_{rad,i}(i=1,2,.....,n)$ in eqn.(6.17), so as to achieve a desired radiation pattern with a specific sidelobe level. Thirdly, $\gamma_i(i=1,2,....,n)$ is calculated using eqn.(6.19). γ_i is equal to $G_{H,i}^*$. The insertion length which realises $G_{H,i}^*$ is obtained from the admittance graph made in advance, giving a value of $B_{H,i}^*$. Therefore, from eqn.(6.14) the coupling phase ϕ_i can be calculated.

Finally, the mechanical rotation angle of the i-th array element, $\phi_{mech,i}$, is given by using the relation of $\phi_i + \phi_{mech,i} = 0$. The mechanical rotation of each array element leads to the in-phase condition at the aperture of the array. For example, if $\phi_i = -\delta$ at the i-th array element shown in Fig.6.29, then the mechanical rotation must be made counterclockwise by δ(rad).

In practice we first insert the n-th array element (the last array element), which satisfies $\gamma_n = G_{H,n}^*$, into the waveguide. After rotating the n-th array element by $\phi_{mech,n}$, we tune it by the n-th stub. Subsequently we insert the (n-1)-th array element, which satisfies $\gamma_{n-1} = G_{H,n-1}^*$, with rotation angle of $\phi_{mech,n-1}$ into the waveguide and tune it by the (n-1)-th stub. Similarly, the insertion, rotation and tuning are done one by one for the (n-2)-th, the (n-3)-th,..., and the first array elements.

6.8.2. Experimental results

The operating frequency is chosen to be 9.375 GHz, which is a typical frequency in the X band, and a waveguide of WRJ-10 is used to feed the array of 1.5-turn helical antennas. The parameters of the helical antenna must be chosen so that a circularly polarised beam may be obtained. The circumference of the helical cylinder is $C=1\lambda$ (λ is free-space wavelength) = 3.2cm. The pitch angle and the wire radius are $\alpha=12.5°$ and $\rho = 0.009375\lambda = 0.3$mm, respectively.

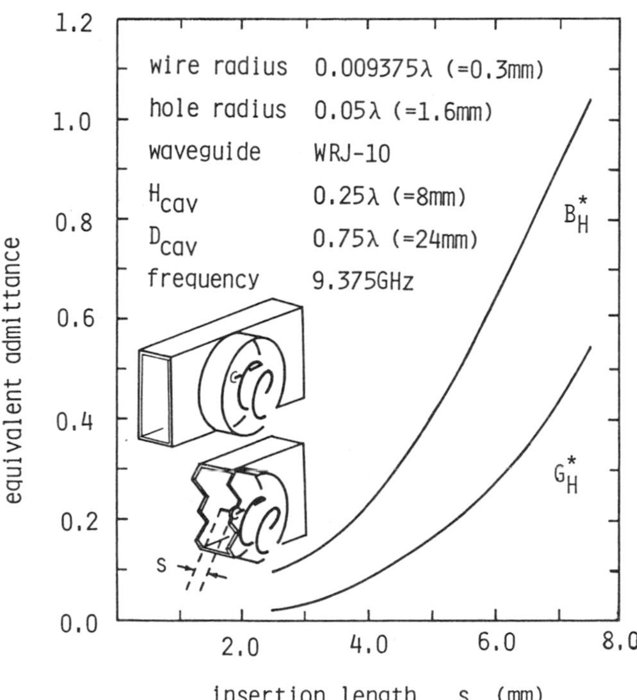

Fig.6.31 Experimental result of equivalent admittance of helical array element with a cavity, $G_H^* + jB_H^*$, versus the insertion length.

Fig.6.31 shows an experimental result of equivalent admittance of the helical array element with a cavity, $G_H^* + jB_H^*$, versus the insertion length. The diameter and height of the cavity are $D_{cav} = 0.75\lambda$ and $H_{cav} = 0.25\lambda$, respectively. The axial ratio is 1.7 dB regardless of the insertion length.

Fig.6.32 shows experimental radiation patterns of a single array element with the mechanical rotation of ϕ_{mech}. The patterns are measured using a circularly polarised antenna as a receiving antenna. It is seen that the radiation pattern remains nearly constant regardless of the mechanical rotation, and is almost symmetrical with respect to the helical axis. Hence, we assume a dotted

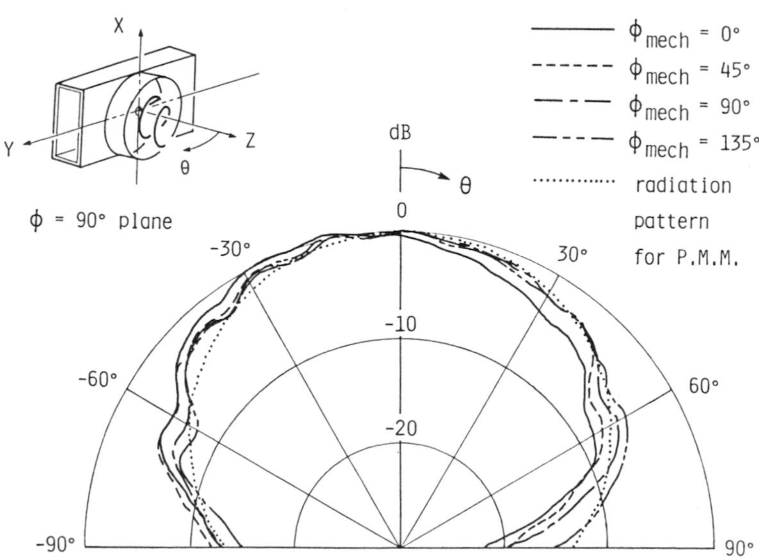

Fig.6.32 Experimental radiation pattern of a single array element with the mechanical rotation of ϕ_{mech}.

line as a radiation pattern of the single array element to calculate the radiation pattern of the array by a pattern multiplication method(PMM).

We show an example of the array design in the case of a power efficiency η of 80 % and CHEBȲSHEV distribution of the sidelobe level of -20 dB in isotropic sources. When the number of the array elements is five, the distance between the array elements is 0.78 λ.

First, the fifth array element is inserted into the waveguide and tuned by a stub. A standing wave ratio meter is used to check the tuning condition. Subsequently, the insertion and tuning are applied in turn to the fourth, the third, the second, and the first array elements.

The radiation patterns of a five-element array without mechanical rotation and then with mechanical rotation are shown in Figs.6.33(a) and (b), respectively. As expected, a main beam is formed by the mechanical rotation of the array elements. The half-power beamwidth (HPBW) in the ϕ=90° plane is 15° (calculated value=15°), and the sidelobe level is -20 dB (calculated value=-21 dB). The axial ratio is 1.7 dB, corresponding to a cross polarisation of -20 dB.

When the number of the array elements is seven, the distance between the array elements is 0.85 λ. For nine elements the distance between the array elements is 0.88 λ. Figs.6.34(a) and (b) show the radiation patterns of the arrays using seven and nine elements, respectively. The HPBW in ϕ=90° plane is related to the change in the number of the array elements. The experimental HPBW's are 9° and 7° for seven and nine elements, respectively. They are in good agreement with the theoretical values.

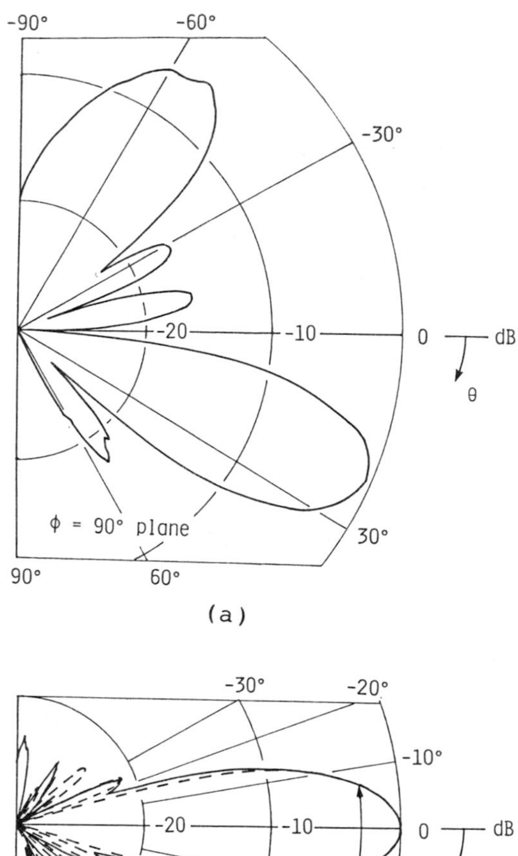

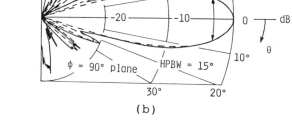

Fig.6.33 Radiation patterns of the array.
(a) Without mechanical rotation.
(b) With mechanical rotation.
The number of helical antennas is five.
– – – theoretical; ——— experimental

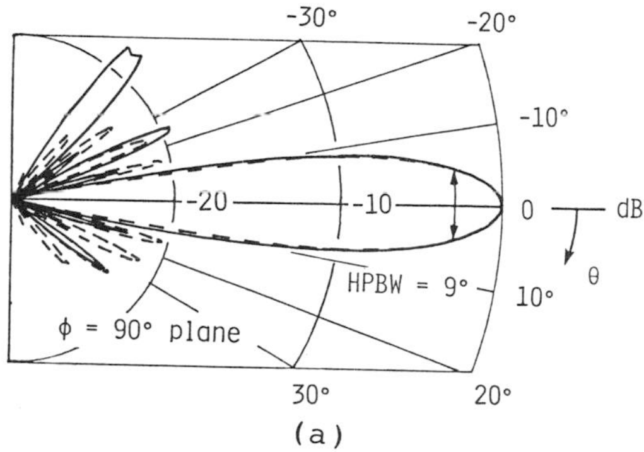

(a)

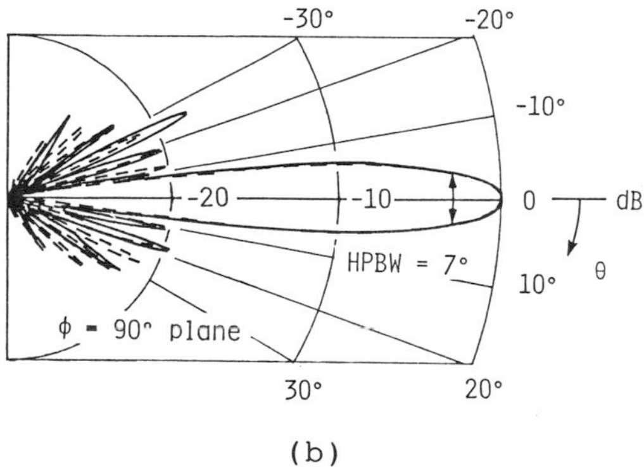

(b)

Fig.6.34 Radiation patterns of the arrays.

(a) Seven helical antennas.

(b) Nine helical antennas.

– – – theoretical; ——— experimental

REFERENCES TO CHAPTER 6

[1] Kraus, J.D., "Antennas", McGraw-Hill, 1950, Chap.7.

[2] King, H.E. and Wong, J.L., "Characteristics of 1 to 8 wavelength uniform helical antennas", IEEE Trans., AP-28, 1980, pp.291-296.

[3] Angelakos, D.J. and Kajfez, D., "Modifications on the axial-mode helical antenna", Proc.IEEE, 55, 1967, pp.558-559.

[4] Wong, J.L. and King, H.E., "Broadband quasi-taper helical antennas", IEEE Trans., AP-27, 1979, pp.72-78.

[5] Donn, C., "A new helical antenna design for better on- and off- boresight axial ratio performance", IEEE Trans., AP-28, 1980, pp.264-267.

[6] Nakano, H. and Yamauchi, J., "Balanced-feed axial-mode helical antenna", TGAP78-19, 1978, pp.23-28.

[7] Nakano, H. and Yamauchi, J., "The balanced helices radiating in the axial mode ", IEEE Proc. of Int. Symp., 1979, pp.404-407.

[8] Shiokawa, T. and Karasawa, Y., "Radiation characteristics of axial-mode helical antenna", IECE, Vol.63-B, 1980, pp.143-150.

[9] Nakano, H. and Yamauchi, J., "Axial ratio of balanced helical antenna and ellipticity measurement of incident wave", Electron. Lett., Vol.17, No.11, 1981, pp.365-366.

[10] Nakano, H. and Yamauchi, J., "Balanced-feed axial-mode helical antenna [II]", TGAP78-74, 1978, pp.25-30.

[11] Nakano, H. and Yamauchi, J., "Radiation characteristics of helix antenna with parasitic elements", Electron. Lett., Vol.16, No.18, 1980, pp.687-688.

[12] Yamauchi, J., Nakano, H. and Mimaki, H., "Balanced helical antenna with tapered open ends", Trans. IECE, Vol.J64-B, No.4, 1981, pp.279-286.

[13] Nakano, H. and Yamauchi, J., "Characteristics of modified spiral and helical antennas", IEE Proc. Vol.129, pt.H, No.5, 1982, pp.232-237.

[14] Nakano, H., Asaka, N. and Yamauchi, J., "Radiation characteristics of short helical antenna and its mutual coupling", Electron. Lett., Vol.20, 1984, pp.202-204.

[15] Stratori, A.R. and Wilkinson, E.J., "An investigation of the complex mutual impedance between short helical array elements", IRE Trans., AP-7, 1959, pp.279-280.

[16] Nakano, H., Asaka, N. and Yamauchi, J., "Short helical antenna array fed from a waveguide", IEEE Trans. AP-32, 1984, pp.836-840.

[17] Agrawal, V.D. and Wong, G.G., "A high performance helical element for multiple access array on TDRSS spacecraft", IEEE Int. Antennas Propagat. Soc. Symp. Digest, 1979, pp.481-484.

[18] Shiokawa, T. and Karasawa, Y., "Array antenna composed of 4 short axial-mode helical antennas", Trans. IECE, Vol. J65-B, No.10, 1982, pp.1267-1274.

[19] King, H.E. and Wong, J.L., "240-400 MHz antenna system for the fleetsatcom satellites", Proc. IEEE International symposium on antennas and propagation, 1977, pp.349-352.

CHAPTER 7
Helical Antenna of Backfire Mode*

7.1. INTRODUCTION

In Chapter 6 we have discussed the radiation characteristics of a balanced helical antenna, which corresponds to a monofilar helix mounted on an infinite ground plane. The numerical results concerning modifications to a helical structure have revealed that tapering the open end greatly contributes to the improvement of the axial ratio.

Although the axial mode of operation is the most widely used, another interesting mode of operation has been reported by Patton [1]. By using a bifilar helical antenna BH, he found a backfire mode in which a circularly polarised wave was radiated in the backward endfire direction. It should be noted that the BH has an advantage over a conventional axial-mode helix in that a ground plane is not needed. However, the front-to-back (F/B) ratio of the BH is not necessarily large (it is about 10 dB). This causes an interference by an incident wave from the undesired direction when the BH is used as a receiving antenna.

In this chapter, we study the radiation characteristics of a backfire helical antenna and a means of improving the F/B ratio. First, we concentrate our attention on the modification of the BH. The radiation characteristics of

*From papers (p-9),(p-12),(p-19),(c-10) and (c-11) listed at the end of this monograph.

a backfire helix with tapered feed end TBH are investigated and compared with those of a conventional BH and a conical bifilar helix CBH. Secondly, the effects of terminating the arm ends are studied over a frequency range of 1:1.7. Information on the location of the phase centre is also presented which is useful when the antenna is employed as a primary feed in a paraboloidal reflector.

7.2. BACKFIRE BIFILAR HELICAL ANTENNA WITH TAPERED FEED END [2]

7.2.1. Configuration

Fig.7.1(a) shows the configuration and co-ordinate system of a TBH which consists of tapered and uniform sections. The tapered section of the TBH is specified by $r=r_0\exp(a\phi_c)$ and the geometric parameters are chosen as follows: generating line of cone $r_0=3.8$ cm; spiral constant $a=0.048$/rad; tapering cone angle $\theta_0=12.5°$. The circumference of the helical cylinder in the uniform section is $\bar{c}=10$ cm. In both sections, the pitch angle and the wire radius are chosen to be $\alpha=12.5°$ and $\rho=0.5$ mm, respectively. To feed the antenna, a straight wire is inserted between two points which are at the beginning of the tapered section and a delta function generator is located at the centre of the straight wire.

For comparison with the TBH, the BH and the CBH shown in Figs.7.1(b) and (c) are also analysed and discussed. The BH is identical to the uniform section of the TBH except for the number of helical turns. The configuration of the CBH is an extension of the tapered section of the TBH. The condition used for comparison is that the three types of bifilar helices have almost the same axial length and arm length.

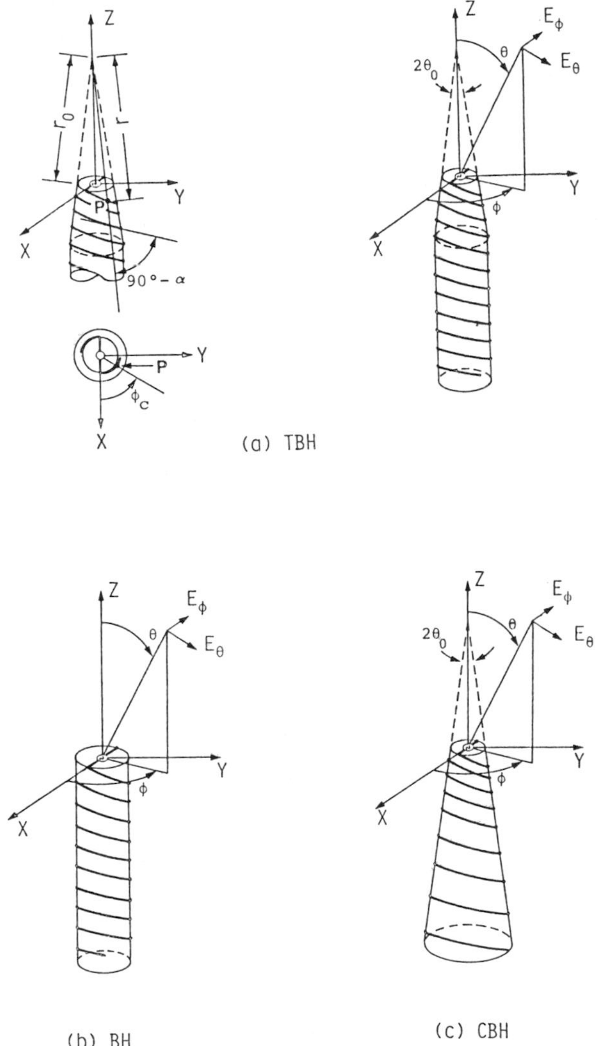

Fig.7.1 Configuration and co-ordinate system.

We choose the helical turns in the uniform section and the tapered section of the TBH to be $n_u=3$ and $n_t=2.2$, respectively. Such precision of the n_t is not necessarily required in practical use. Auxiliary calculation shows that remarkable improvement of the F/B ratio can be obtained provided that the n_t is more than 2 turns. It should also be noted in the present comparison that the TBH occupies the smallest space, which is nearly half that of the CBH.

The analysis is carried out over a frequency range of 2.0 GHz to 3.5 GHz which gives a corresponding circumference range of 0.67 λ to 1.17 λ (λ=free-space wavelength). This is based on Patton's experimental results [1], which show that a BH with a pitch angle of about 12° is expected to operate in the backfire mode, when the circumference is in a range of 0.7 λ to 1.1 λ.

7.2.2. Current distribution and input impedance

Fig.7.2(a) shows the typical current distribution of the TBH with a source of $V_0=1$ volt at 3.0 GHz. The behaviour of the current amplitude as a function of frequency is shown in Fig.7.2(b). As the frequency is increased, the current decays smoothly along the arm and the attenuation rate becomes larger. At 2.3 GHz, the amplitude at the end point of the tapered section is reduced to about 5 dB below the maximum, while at 3.5 GHz, about 17 dB below the maximum. It is also noted from Fig.7.2(b) that a standing wave becomes noticeable at 2.1 GHz. This means an increase in the reflected current from the arm end and deteriorates the F/B ratio, as will be mentioned in the Section 7.2.3. The current phase in Fig.7.2(a) shows that the phase progression along the arm is nearly equal to the free-space phase progression, particularly in the tapered

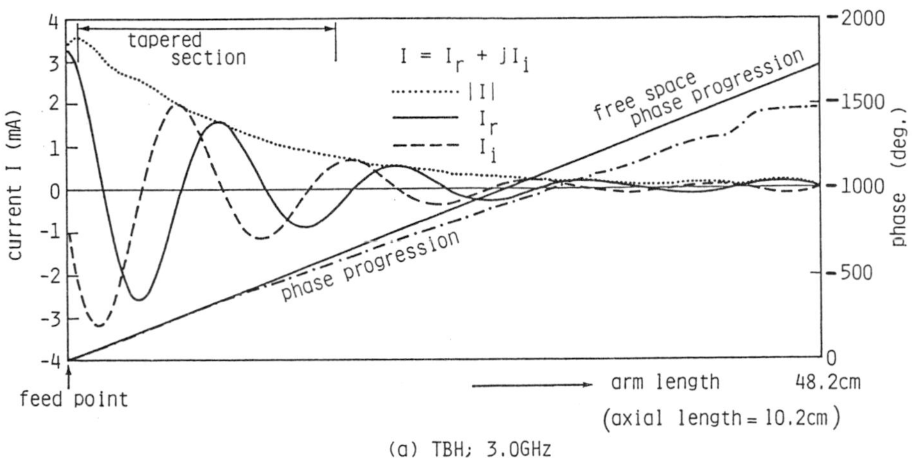

(a) TBH; 3.0GHz

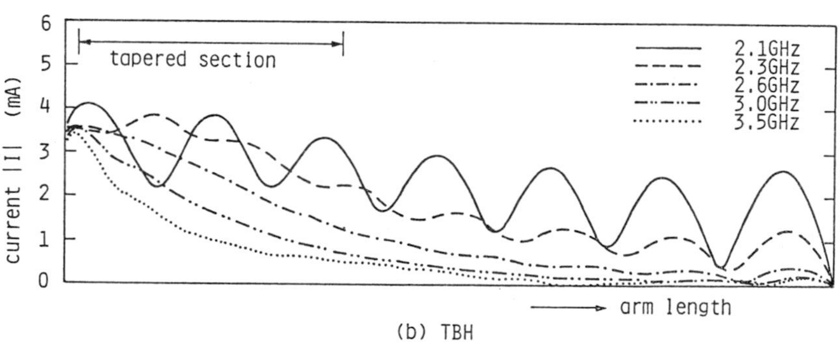

(b) TBH

Fig.7.2(a),(b) Current distributions.

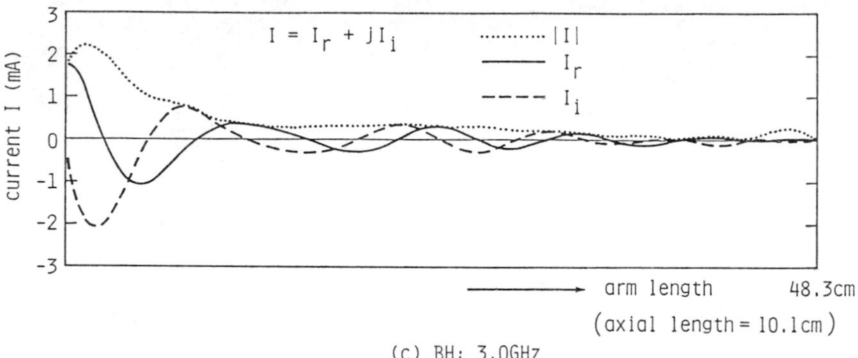

(c) BH; 3.0GHz

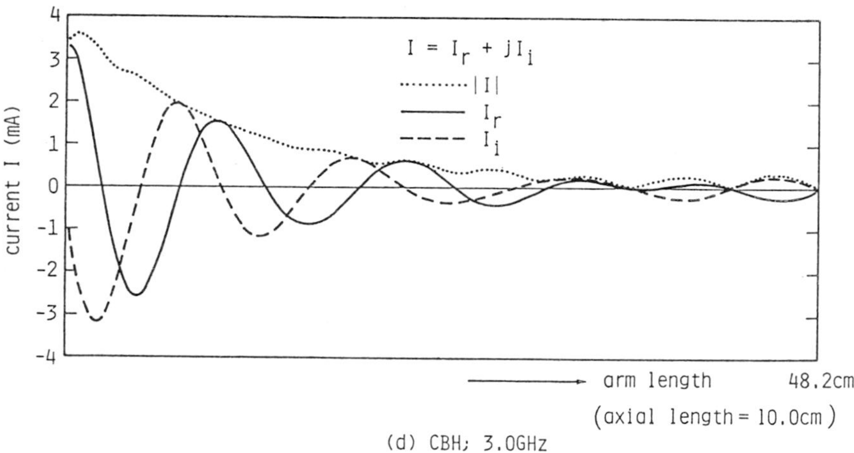

(d) CBH; 3.0GHz

Fig.7.2(c),(d) Current distributions.

section. Further calculation supports this tendency over a frequency range of 2.3 GHz to 3.5 GHz. The current behaviour of the TBH contrasts with that of a conventional axial mode helix in which a slow wave with constant amplitude is generated.

The current distributions of the BH and the CBH at 3.0 GHz are also shown in Figs.7.2(c) and (d). It is observed that the attenuation of the current of the BH is not necessarily smooth as in the case of the TBH, and that the value of the current at the input is considerably different from that of the TBH. In contrast, we find a similarity between the current attenuations on the TBH and the CBH. This is due to the fact that the tapered section of the TBH is the same as the beginning part of the CBH, and that a considerable contribution to the radiation occurs from this tapered section.

Although the attenuation rate of the current of the TBH is changed when the frequency is varied, the values of the current at the input remain relatively constant. This leads to a wideband characteristic of the input impedance not obtainable in the conventional BH, as shown in Fig.7.3. In the present model, the resistance value is of the order of 300 ohms and the reactance value is of the order of 40 ohms over a frequency range of 2.3 GHz to 3.5 GHz. In addition, it should be noted that the input impedance of the TBH at frequencies higher than 2.3 GHz is almost the same as that of the CBH because of the similarity between the current distributions.

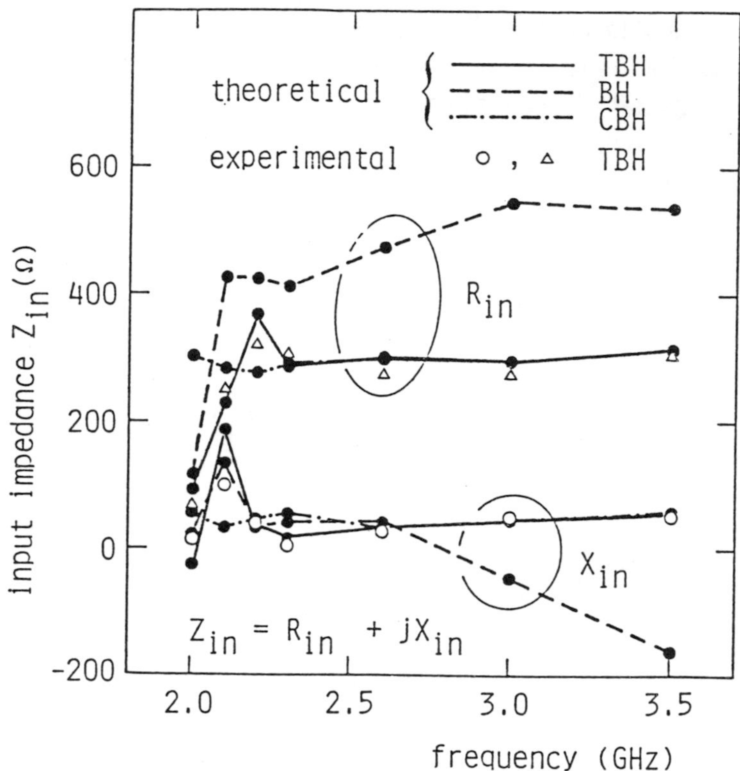

Fig.7.3 Input impedance.

7.2.3. Radiation characteristics

The typical radiation pattern of each helix at 3.0 GHz is shown in Fig.7.4. The radiation patterns are shown only for the $\phi=0°$ plane, since they have relatively good

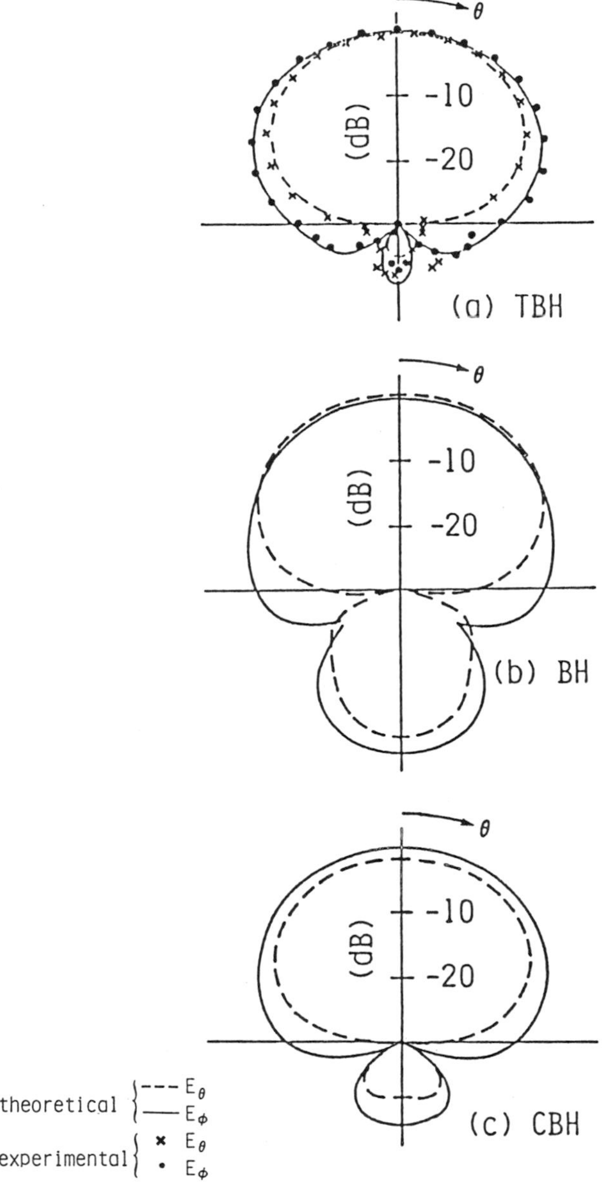

Fig.7.4 Radiation patterns; 3.0 GHz, ϕ=0° plane.

similarity in the azimuth. Each pattern is essentially backfire, but we find a difference in the level of the back lobe which occurs in the direction opposite to the main lobe.

Fig.7.5 shows the average F/B ratio of the E_θ and E_ϕ components as a function of frequency. By tapering only the feed end, remarkable improvement can be obtained in the F/B ratio. For example, at 3.0 GHz the F/B ratio of the BH is 6 dB, while that of the TBH is 23 dB. It is interesting to note that the F/B ratio of the TBH is also larger than that of the CBH over a frequency range of 2.6 GHz to 3.0 GHz. In the vicinity of 2.0 GHz, the F/B ratios of the TBH and the BH tend to be small. This is due to an increase in the reflected

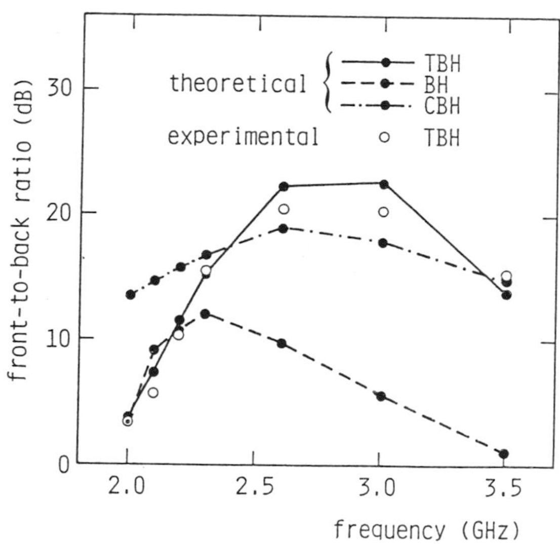

Fig.7.5 Front-to-back ratio.

current, which is responsible for the radiation toward the -Z-axis. Since the current distributions of the TBH and the BH become similar as the frequency is decreased, the noticeable difference in the F/B ratio between these two antennas does not appear at frequencies lower than 2.2 GHz.

The average half-power beamwidth (HPBW) of the E_θ and E_ϕ components of the main lobe is shown in Fig.7.6. In general, the HPBW of the TBH is narrower than those of the BH and the CBH. It is also found that the HPBW's of the TBH and the BH decrease with a decrease in frequency.

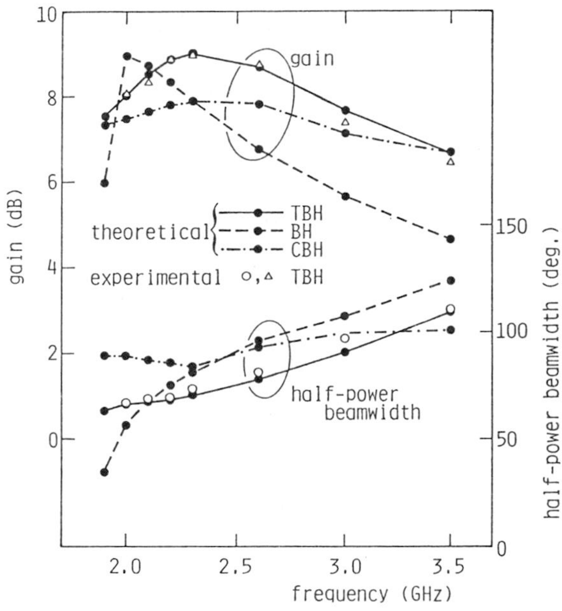

Fig.7.6 Half-power beamwidth and power gain.

The behaviour of the HPBW is closely related to the change in the gain. Fig.7.6 also shows the power gain characteristics. Over a frequency range of 2.2 GHz to 3.0 GHz, the gain of the TBH is remarkably enhanced as compared with those of the other helices, although the occupied space of the TBH is the smallest. This fact is attributed to the narrowness of the HPBW. In a range of this numerical result, a maximum gain of 9.0 dB is obtained at 2.3 GHz. It is also seen from Fig.7.6 that the gain of the CBH remains relatively constant in spite of the change in frequency.

Fig.7.7 shows the axial ratio. It is revealed that in comparison with the BH, the TBH has an excellent axial ratio at frequencies higher than 2.6 GHz because of the smooth decay in the current distribution, as shown in Fig.7.2(b). The axial ratio of the TBH is less than 1 dB over a frequency range of 2.3 GHz to 3.0 GHz.

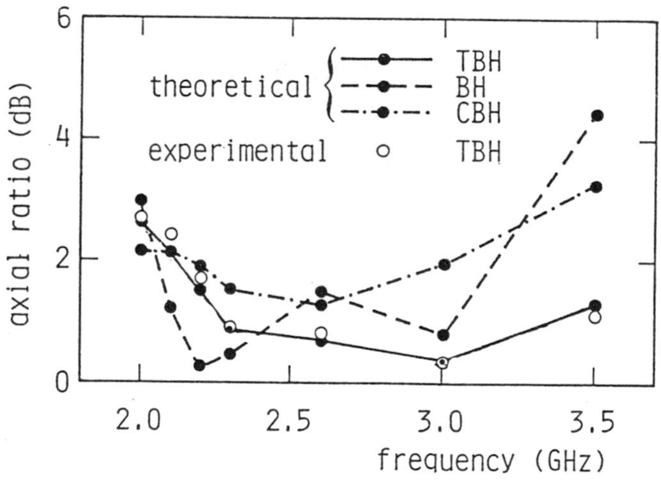

Fig.7.7 Axial ratio.

7.3. TAPERED BACKFIRE HELICAL ANTENNA WITH LOADED TERMINATION [3]

The present section is devoted to a comprehensive study of a tapered backfire helical antenna with loaded termination LTBH shown in Fig.7.8. Taking account of the value of loaded resistance R_L, we determine the current distribution along the arm. The characteristics of the LTBH, including the input impedance, radiation pattern, axial ratio and power gain, are calculated over a frequency range of 2.1 to 3.5 GHz and compared with those of a backfire helix with tapered feed end TBH without loaded termination.

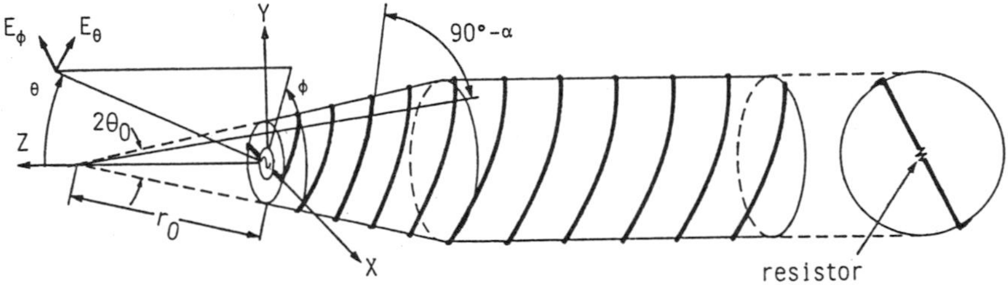

Fig.7.8 A tapered backfire helical antenna with loaded termination LTBH

generating line of conical r_0=3.8 cm

spiral constant a=0.048/rad

pitch angle α=12.5°

tapering cone angle θ_0=12.5°

wire radius ρ=0.5 mm

helical turns in tapered section n_t=2.2

helical turns in uniform section n_u=3

7.3.1. Determination of the value of loaded resistance

According to Section 7.2.2, the current of a TBH attenuates smoothly along the arm, and the attenuation rate of the current decreases with a decrease in frequency. Hence, the effects of loaded termination of an LTBH should be larger at lower frequencies. The determination of the value of loaded resistance is made at a lower operating frequency of 2.3 GHz. The numerical treatment of loaded termination is found in Reference [3].

Fig.7.9 shows the F/B ratio and the axial ratio as functions of loaded resistance R_L. The F/B ratio is shown by the average of those of the E_ϕ- and E_θ- components. The LTBH realises an F/B ratio of more than 20 dB and an

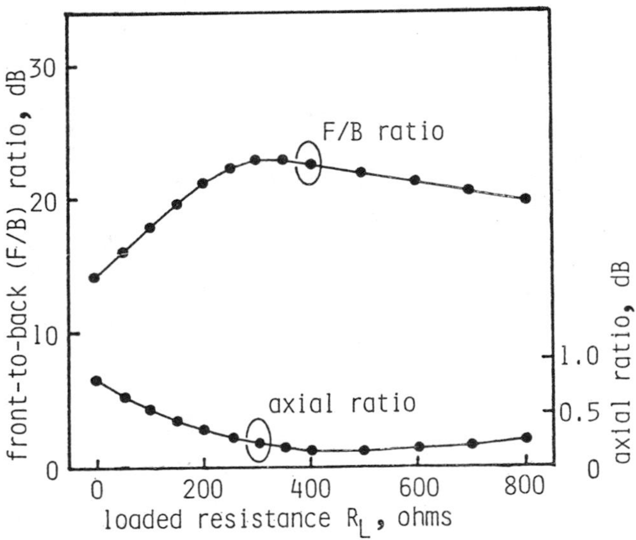

Fig.7.9 F/B ratio and axial ratio as functions of loaded resistance R_L at 2.3 GHz.

axial ratio of less than 0.5 dB over a resistance range of 150 to 800 ohms. A maximum F/B ratio of 23 dB is obtained at 300 ohms, and a minimum axial ratio of 0.2 dB is obtained at 500 ohms.

7.3.2. Frequency characteristics

Based on the results mentioned in Section 7.3.1, the characteristics of the LTBH with R_L=300 ohms are investigated in detail over a frequency range of 2.1 to 3.5 GHz. The resistor used in the experiment is a film type resistor making use of surface resistivity. The resistor is so small (of the order of 1×2 mm^2) that it is regarded as a lumped resistance. For comparison, the characteristics of the TBH without loaded termination are also presented. The only difference in geometry between the LTBH and the TBH is that straight wires connected to a resistor are inserted between the ends of the uniform section in the LTBH, as shown in Fig.7.8.

Fig.7.10 shows the typical current distribution of the LTBH at 2.2 GHz. As additional information, the current of the TBH is also presented. It is found that the noticeable standing wave observed in the TBH does not appear in the LTBH, since the reflected current is reduced by the terminal resistor. If the arm end is not terminated with the resistor, a greater length of antenna is required for the establishment of a travelling current distribution throughout the arm. The use of terminal resistor makes it possible to yield the smooth attenuation of the current distribution, even when the antenna length is limited. Fig.7.10 also shows the phase progression of the LTBH. The phase changes linearly along the arm because of the establishment of an outgoing travelling wave toward the arm end.

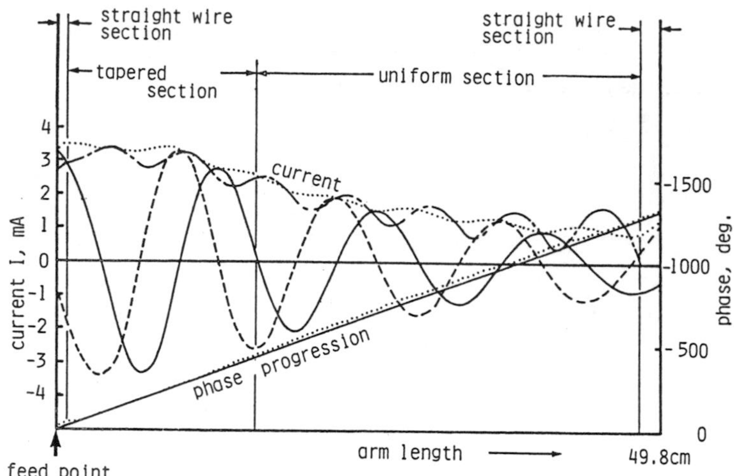

Fig.7.10 Current distribution and phase progression at 2.2 GHz.

Current $I=I_r+jI_i$

LTBH: ••••••• $|I|$ ——— I_r - - - - I_i

TBH: —— - —— $|I|$

Phase progression

••••••• LTBH ——— free space

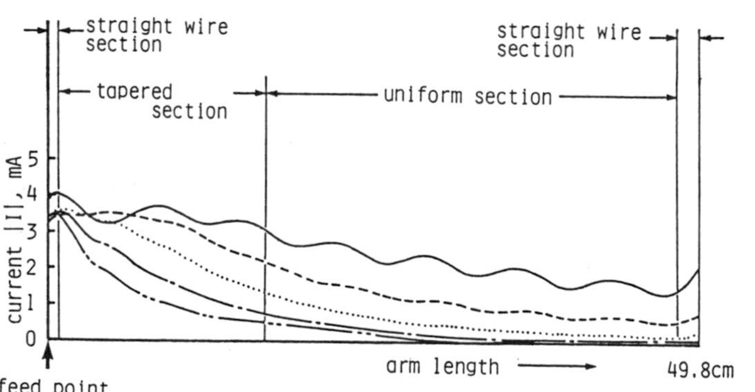

Fig.7.11 Current distribution as a function of frequency.

——— 2.1 GHz — — — 2.3 GHz •••••• 2.6 GHz

—·— 3.0 GHz —··— 3.5 GHz

The behaviour of the current amplitude as a function of frequency is shown in Fig.7.11. It can be seen that the current attenuates smoothly along the arm, even when the frequency is changed. At 3.0 and 3.5 GHz, the current substantially attenuates before it reaches the arm end. This leads to the fact that the radiation characteristics of the LTBH become nearly the same as those of the TBH at higher operating frequencies, as will be mentioned later.

Since the value of the current at the input remains relatively constant in spite of the change in frequency, a wideband characteristic of the input impedance is realised, as shown in Fig.7.12. In the present LTBH, the input impedance values are about 300 ohms resistive and about 40 ohms reactive at frequencies from 2.1 to 3.5 GHz. It should be noted that the variation in the input

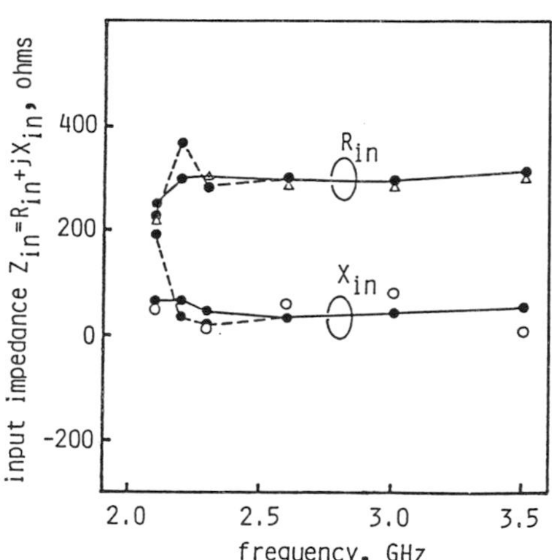

Fig.7.12 Input impedance.

Theoretical: ———— LTBH – – – – TBH

Experimental: △ ○ LTBH

impedance of the TBH, which is observed at lower operating frequencies, is almost eliminated in the LTBH.

Fig.7.13(a) shows the radiation patterns of the LTBH when the frequency is changed. The experimental results are obtained by feeding the antenna from a semirigid coaxial line of 2.19 mm outer diameter. The semirigid coaxial line with a split-tube balun is located on the -Z-axis in Fig.7.8. For comparison, the calculated radiation patterns of the TBH are also presented in Fig.7.13(b). It should be noted that at frequencies lower than 2.6 GHz the level of the backlobe of the LTBH is markedly reduced, as compared with that of the TBH. For example, at 2.1 GHz an average F/B ratio of 7 dB for the TBH is improved to 18 dB for the LTBH. At 3.5 GHz the level of the backlobe of the LTBH is almost the same as that of the TBH because of the similarity between the current distributions of these two antennas. A comparison of radiation patterns in Fig.7.13 also shows that there is essentially no difference in the shape of the main beam between the LTBH and the TBH.

Fig.7.14 shows the axial ratio as a function of frequency. The loaded termination contributes to the improvement of the axial ratio, especially at lower operating frequencies. For example, at 2.1 GHz, the axial ratio of the LTBH is 0.8 dB, while that of TBH is 2.2 dB. Thus the LTBH realises a good axial ratio of less than 1 dB over a frequency range of 2.1 to 3.5 GHz. Further calculation shows that the axial ratio within 3 dB is realised over a wide angle of more than $\theta=\pm55°$ over the same frequency range.

Fig.7.14 also shows the power gain characteristic. Although the use of terminal resistor reduces the radiation efficiency, the gain of LTBH is nearly the same as that of TBH, as shown in Fig.7.14, because the power dissipated in the resistor is compensated for by the

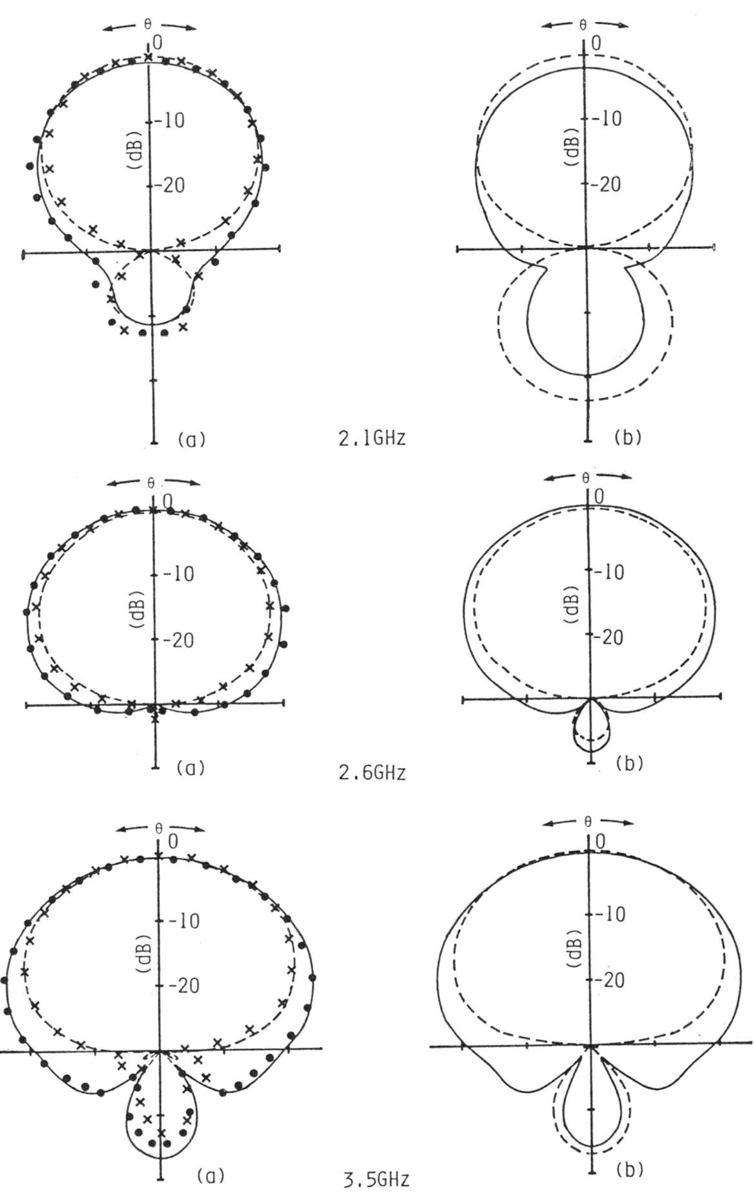

Fig.7.13 Radiation patterns (ϕ=0° plane); (a):LTBH (b):TBH

Theoretical: ——— E_ϕ - - - - E_θ

Experimental(LTBH): ● E_ϕ ✖ E_θ

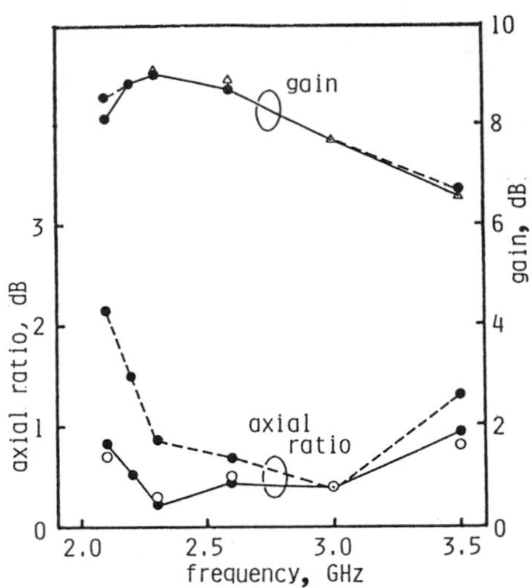

Fig.7.14 Axial ratio and gain.
Theoretical: ———— LTBH - - - - TBH
Experimental: △ ○ LTBH

improved directivity which results from the loss of radiation in the forward endfire direction.

The maximum gain of the LTBH is obtained at a frequency of 2.3 GHz, which gives a circumference of 0.77 λ in the uniform helical section. It is interesting to note that the circumference of 0.77 λ corresponds to a circumference in which a conventional axial-mode helix with a ground plane is just about to establish the axial mode radiation. The conventional axial-mode helix requires a circumference of about 1 λ when it satisfactorily operates as a circularly polarised antenna. In other words, the LTBH has an advantage in that the cross-section of helical cylinder in terms of the wavelength is smaller than that of the conventional axial-mode helix.

7.3.3. Phase centre

It has been found that the LTBH realises a good F/B ratio over a wide frequency range, with smaller occupied space as compared with that of a conventional axial-mode helix mounted on a ground plane. Thus it is expected that the LTBH can be used as a primary feed for a paraboloidal reflector. To perform as an efficient feed for the reflector, it is imperative that the phase centre of the LTBH be known.

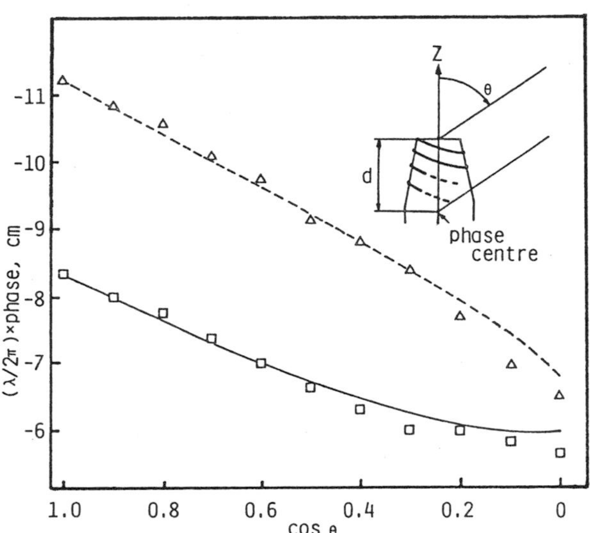

Fig.7.15 Phase change as a function of cosθ at 2.6 GHz $\phi=0°$ plane

Theoretical: ——— E_ϕ - - - - E_θ

Experimental: □ E_ϕ △ E_θ

The phase centre is found by plotting the phase change of the radiation field against cosθ [4]. The distance d from the origin to the phase centre, shown in the inset of Fig.7.15, is the slope of the linear portion of the plot. The plot will be linear only if the wavefronts are truly spherical. For a practical antenna, the plot is linear only within a certain angular region. Within that angular region, the field is spherical and the location of the phase centre is the slope of the linear portion. If no part of the plot is linear, the assumed discrete phase centre does not exist.

As a typical example, the phase change of the LTBH at 2.6 GHz is shown in Fig.7.15. The phase changes linearly for both E_ϕ- and E_θ- components over a wide angle of $\theta=\pm70°$($\cos\theta=0.34$). The calculated results are in close agreement with the experimental results.

Fig.7.16 shows the phase centre location d for the E_θ- and E_ϕ- components as a function of frequency. The data are presented for both $\phi=0°$ and $\phi=90°$ planes. When the LTBH is used for a primary feed of a circularly polarised wave, it is desirable that four curves in Fig.7.16 coincide with one another. Although these curves are not strictly identical in the present model, a maximum difference among these curves is only 0.05 λ. It is also found from Fig.7.16 that as the frequency is increased the phase centre location of the LTBH moves toward the feed point. This is consistent with the fact that, as mentioned in Section 7.2.2, the attenuation rate of the current increases with an increase in frequency, and the active region responsible for the radiation moves toward the feed point.

In summary, the LTBH has an apparent phase centre over a wide frequency range of 2.1 to 3.5 GHz.

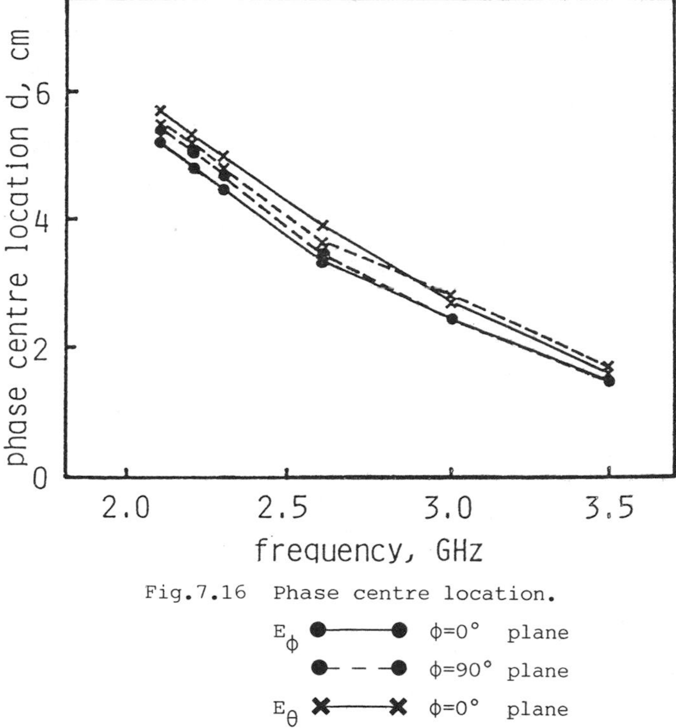

Fig.7.16 Phase centre location.

7.4. CONCLUSION

A backfire bifilar helical antenna with tapered feed end TBH has been numerically and experimentally investigated, and compared with conventional bifilar and conical bifilar helices. It is revealed that the TBH has a higher F/B ratio as compared with the conventional bifilar helix. Tapering only the feed end also improves the axial ratio particularly at higher frequencies. The gain of the TBH is higher than those of the other helices, with the advantage that the TBH occupies only half the space of the conical bifilar helix.

Subsequently the radiation characteristics of a tapered backfire helical antenna with loaded termination (LTBH) have been investigated. The loaded resistance R_L is typically chosen to be 300 ohms. It is demonstrated that terminating the arm ends with a resistor leads to a wide-band characteristic of the input impedance and improves the front-to-back ratio and the axial ratio at lower operating frequencies. The inherent characteristics of the power gain and the half-power beamwidth of a tapered backfire helix are not significantly affected by the use of the terminal resistor.

Further calculation reveals that the LTBH has an apparent phase centre, and that the location of the phase centre for the E_ϕ- component is nearly the same as that for the E_θ- component. The LTBH may be used as a primary feed for a palaboloidal reflector, with the advantage that the space occupied by the LTBH is smaller than that of a conventional axial-mode helix with a ground plane.

REFERENCES TO CHAPTER 7

[1] Patton, W.T., "The backfire bifilar helical antenna", University of Illinois, Antenna Laboratory Technical report 61, 1962.

[2] Nakano, H., Yamauchi, J., Mimaki, H. and Iio, S., "Backfire bifilar helical antenna with tapered feed end", Int. J. Electron., vol.54, no.2, 1983, pp.279-286.

[3] Nakano, H., Iio, S. and Yamauchi, J., "Frequency characteristics of tapered backfire helical antenna with loaded termination", Proc. IEE, vol.131, no.3, 1984, pp.147-152.

[4] Hu, Y.Y., "Determination of phase centre of radiating systems", J.Franklin Inst., vol.271, 1961, pp.21-29.

CHAPTER 8
Conical Helix Antenna*

8.1. INTRODUCTION

One of the means of widening the operating bandwidth of a helical antenna of axial mode is to expand the antenna arm into a conical one. An earlier study of the radiation field of a monofilar conical helix antenna with a low pitch angle was made on the assumption that only a progressive current was present along the arm [1]. As a result, it was found that the radiation pattern from the helix was a single lobe with the maximum in the axial direction over a wide range of frequencies.

Recently, a similar configuration was treated from the point of view of lowering the silhouette of the conical helix antenna, and the radiation characteristics were experimentally investigated [2]. It was shown that the performance of the antenna is almost as good as an ordinary conical log-spiral antenna [3] despite the fact that its height is reduced to one-quarter of the corresponding conical log-spiral antenna. The current distribution for the monofilar conical helix, however, has not been obtained by theoretical methods.

The purpose of this chapter is to numerically analyse the monofilar conical helix with a low pitch angle and to discuss the radiation characteristics. Detailed calculations show that there are two distinct regions in

*From papers (p-24),(p-26),and (c-16) listed at the end of this monograph.

the current distribution, and reveal how these two regions contribute to the total radiation field. In Section 8.8 we also study a short conical helix antenna, which has a low silhouette of about 0.3 wavelengths.

8.2. CONFIGURATION

Fig.8.1 shows the configuration and co-ordinate system of a monofilar conical helix. The antenna is mounted on a perfectly conducting plane of infinite extent, allowing the use of image theory in the analysis, as shown in Fig.8.1(c). The antenna arm is defined as an equiangular spiral function $r=r_0\exp(a\phi_c)$. From a practical point of

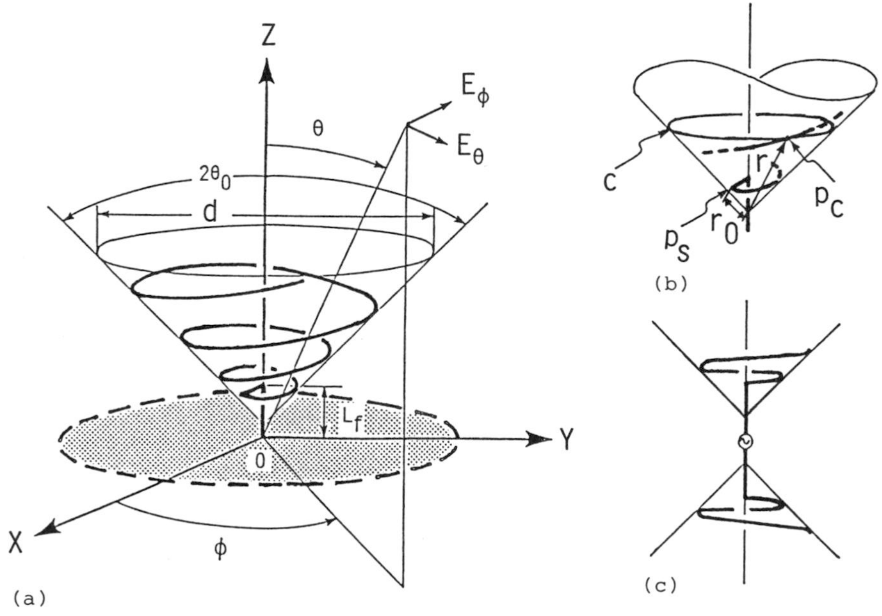

Fig.8.1 Configuration and co-ordinate system.

view, a short wire for the feeding is used at the input. The source is located at the origin O. Referring to the experiment [1], we choose the geometrical parameters as follows: r_0=0.707 cm; a=0.074/rad (giving the pitch angle τ=6°); apex angle of the cone $2\theta_0$=90°; wire radius ρ=0.3 mm; arm length defined as an equiangular spiral function L_e=48.25 cm; diameter of the cone at the arm end d=8.1 cm; feed wire length L_f=1.0 cm.

The analysis is made over a range of frequencies from 1.5 GHz to 6.0 GHz. In the present analysis, the segment length is taken to be 2.5 mm, and the source is treated as a delta-function generator. To verify the numerical results, experimental data are also presented. The experiment uses a balanced-type configuration, which is shown in Fig.8.1(c), instead of using a model with an infinite ground plane. The bazooka balun or the split coaxial balun is used to feed the antenna.

8.3. DETERMINATION OF THE FEEDING WIRE LENGTH [4]

Before considering the characteristics of the monofilar conical helix, we determine the feeding wire length L_f on the Z-axis. The effect of the wire length primarily appears in the input impedance. Fig.8.2 shows the theoretical input impedance as a function of the feeding wire length. The frequency is 3.0 GHz. It is found that, as the feeding wire is lengthened, the resistance increases and the reactance changes in sign from negative to positive. In the present model, a pure resistance of 140 ohms is obtained when the wire length is L_f=1 cm=0.1 λ (λ=free-space wavelength). In the following sections, a monofilar conical helix with a feeding wire of 1 cm is investigated in detail.

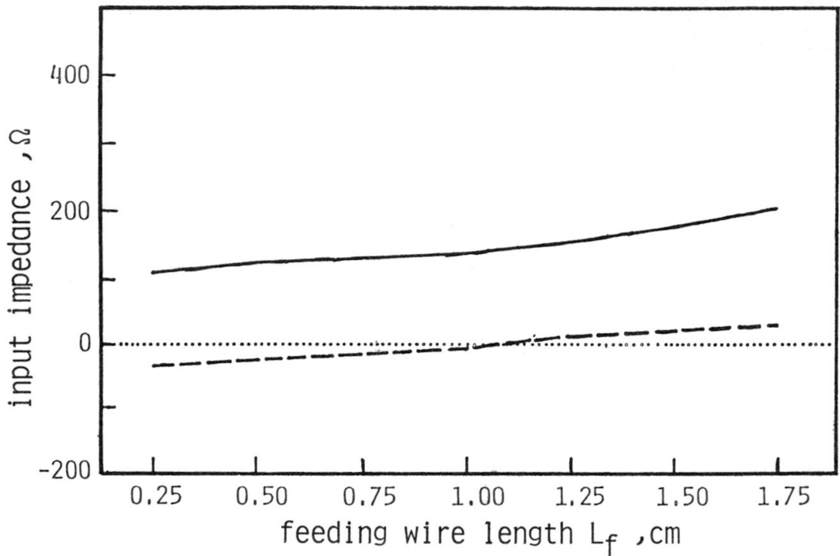

Fig.8.2 Input impedance as a function of the feeding wire length on the Z-axis at 3.0 GHz.

$Z_{in} = R_{in} + jX_{in}$

——— R_{in}

----- X_{in}

8.4. CURRENT DISTRIBUTION AND INPUT IMPEDANCE [4]

Fig.8.3 shows the current distributions at frequencies of 2.5, 3.0, and 4.0 GHz. The applied voltage is 1 volt. The current distribution at 2.5 GHz shows a standing wave, whereas at frequencies higher than 3.0 GHz an outgoing current wave with attenuation appears. It is found that there are two distinct regions in the current distribution: one is a region where the current travels with a phase velocity nearly equal to the velocity of light, decaying from near the source point to the first minimum point, which is indicated by an arrow A_1; the other is a region where the residual current follows after the first minimum point. For convenience, the former is termed region D, and the latter region R.

It should be noted that the behaviour of the current on the monofilar conical helix differs from that on a bifilar conical helix in that the region R does not appear in the bifilar conical helix, as seen in Section 7.2. This is due to the presence of a ground plane. Using the image theory, the monofilar conical helix with a ground plane is equivalent to a balanced conical helix composed of real and image arms which are wound in the axial directions opposite to each other. Calculation shows that the decaying current near the feed point of the monofilar conical helix contributes to the backfire radiation. The radiation from the decaying current on the image arm excites the real arm and induces currents of different modes. Thus the current on the real arm is composed of at least two travelling waves with different modes, resulting in the appearance of the current minima. In other words, the ground plane reflects the radiation from the real arm, and the reflected radiation generates currents of different modes on the real arm.

In the bifilar conical helix without a ground plane,

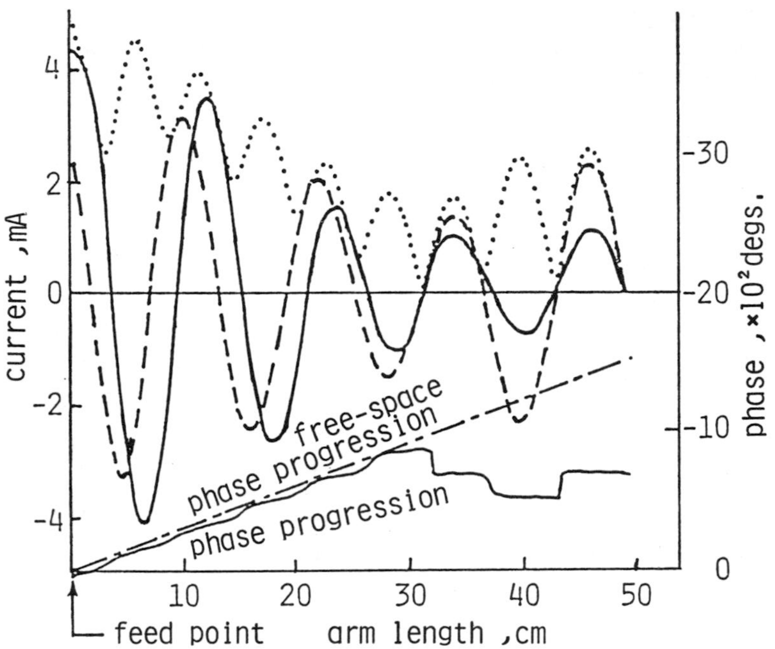

Fig.8.3(a) Current distribution (2.5 GHz).

$I = I_r + jI_i$

.... $|I|$

———— I_r

------ I_i

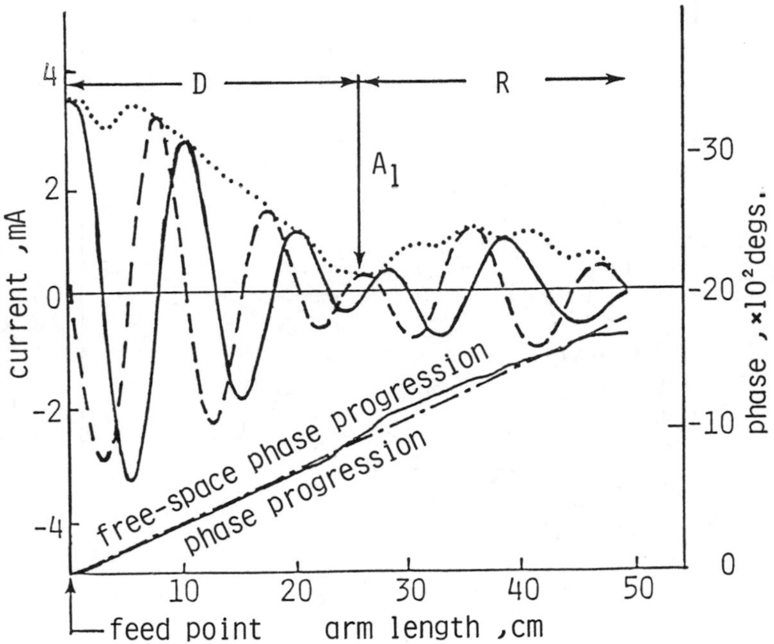

Fig.8.3(b) Current distribution (3.0 GHz).

$I = I_r + jI_i$

.... $|I|$

———— I_r

------ I_i

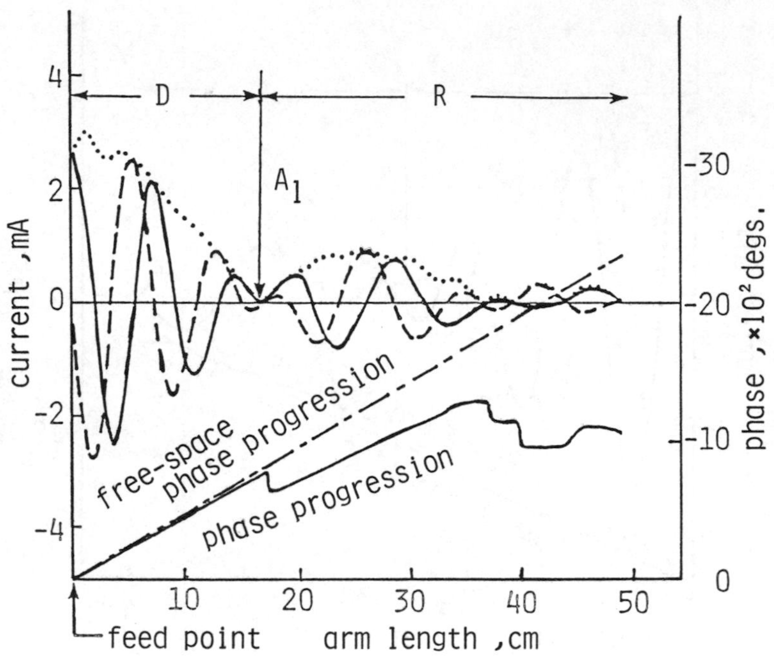

Fig.8.3(c) Current distribution (4.0 GHz).

$I = I_r + jI_i$

.... $|I|$

——— I_r

------ I_i

whose arms are wound in the same axial direction, the backfire radiation due to the decaying current near the feed point is not reflected because of the absence of the ground plane. It follows that no currents are superposed on the decaying current, and the interference of the currents of different modes is not observed.

Fig.8.4 shows the current-amplitude distributions as a function of the arm length expressed by the circumference. A point P_c on the wire occupies a point on the conical surface whose circumference is designated as C, as shown in Fig.8.1(b). The minimum circumference, on which the starting point P_s of the conical wire is located, changes from 0.19 λ to 0.63 λ as the frequency increases from 1.8 to 6.0 GHz. It should be emphasised that, at frequencies higher than 3.0 GHz, the form of the current distribution remains almost unchanged. This leads to the wideband characteristic of the radiation field, as will be mentioned later. The first minimum point, the endpoint of the region D, always exists at a circumference of about 1.3 λ.

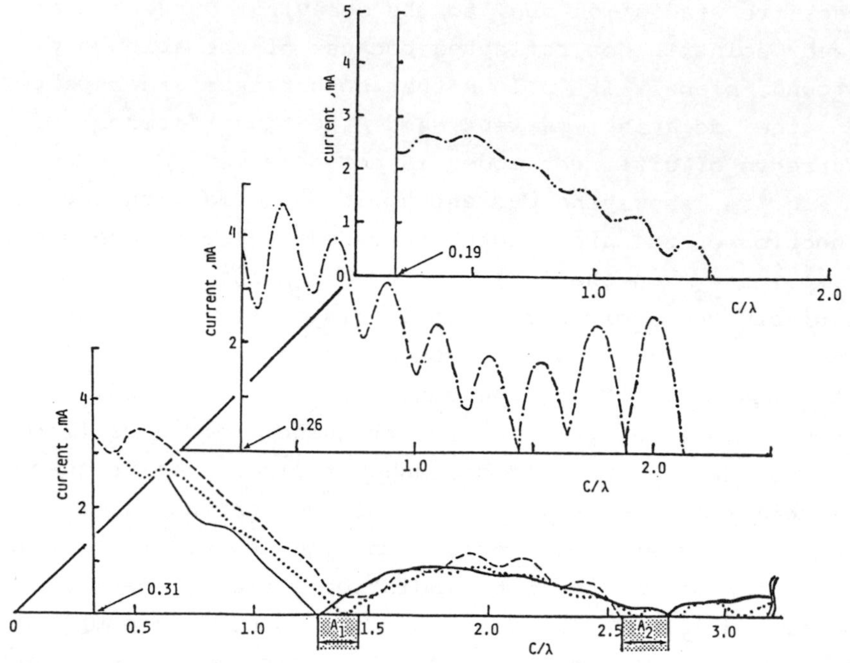

Fig.8.4 Current amplitude distributions.

$$C/\lambda = \frac{2\pi a L_\lambda \sin\theta_0}{\sqrt{a^2+\sin^2\theta_0}} + \frac{2\pi r_0 \sin\theta_0}{\lambda}$$

C = circumference of the cone

L_λ= arm length L_e normalised to the operating wavelength λ

-••-••-•• 1.8 GHz

-•-•-•-• 2.5 GHz

------- 3.0 GHz

•••••••••• 4.0 GHz

——— 6.0 GHz

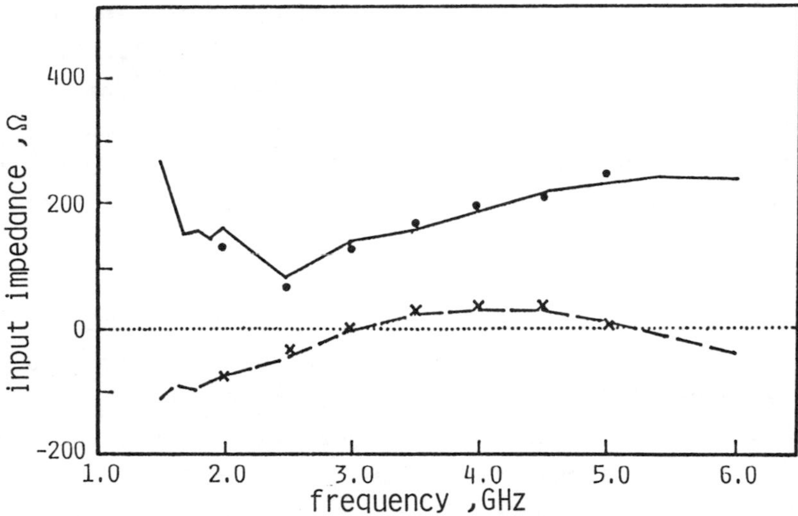

Fig.8.5 Input impedance as a function of frequency.

$Z_{in} = R_{in} + jX_{in}$

Theoretical	Experimental
——— R_{in}	• • • • • • R_{in}
------- X_{in}	× × × × × × X_{in}

Fig.8.5 shows the input impedance as a function of frequency. The input impedance is nearly a pure resistance of the order of 200 ohms over a range of frequencies of 3.0 to 6.0 GHz. Experimental data are also presented. Since the balanced-type configuration is used in the experiment, the half value of the input impedance measured in practice is plotted. The experimental results confirm the wideband characteristic of the input impedance.

8.5. RADIATION CHARACTERISTICS [4]

Fig.8.6 shows typical radiation patterns at 3.0 and 5.0 GHz. It is found that the conical helix gives a relatively symmetrical pattern with respect to the Z-axis, although the configuration is not symmetrical. The radiation pattern has a wideband characteristic and the half-power beamwidth is about 90° at frequencies higher than 3.0 GHz. It is also found that the radiation field is circularly polarised on the Z-axis, as shown in Fig.8.7, where the axial ratio is evaluated in the endfire direction. An axial ratio of less than 1 dB is obtained over a range of frequencies from 3.0 to 5.5 GHz.

Owing to the wideband characteristic of the radiation pattern, it follows that the power gain also has a wideband characteristic, as shown in Fig.8.7. The slight change in the gain is due to the feed structure. To be more precise, as the frequency is increased, the main region of current distribution (region D) shifts toward

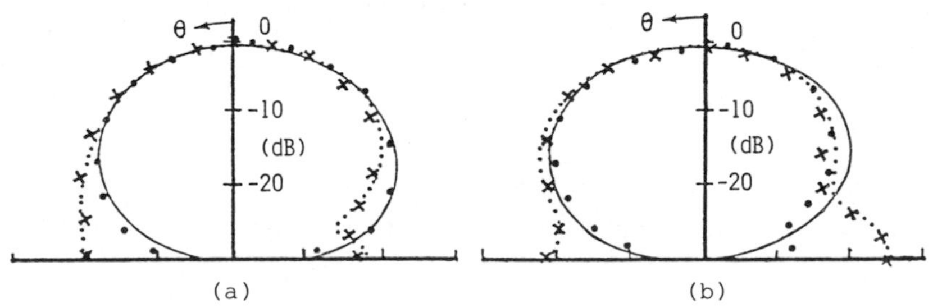

Fig.8.6 Radiation patterns (ϕ = 0° plane).
(a) 3.0 GHz (b) 5.0 GHz
Theoretical Experimental
——— E_ϕ ● ● ● ● ● E_ϕ
·········· E_θ × × × × × E_θ

the feed point and the short wire for the feed begins to act as the radiating element of a linearly polarised wave. In this model, the power gain is 6.7 ± 0.8 dB at frequencies ranging from 3.0 to 6.0 GHz, where a circularly polarised wave is radiated.

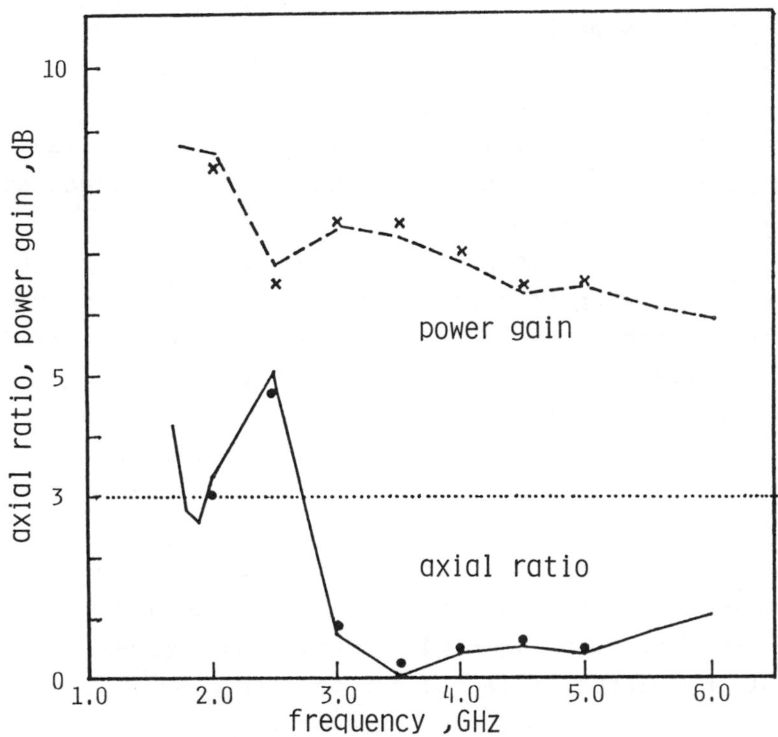

Fig.8.7 Axial ratio and power gain in the endfire direction.

Theoretical	Experimental
——— axial ratio	● ● ● ● axial ratio
------ gain	× × × × gain

8.6. EFFECTS OF REGIONS D AND R ON THE RADIATION CHARACTERISTICS [4]

In order that we may recognise how the two regions D and R are responsible for the radiation field, some observations are made about the radiation field at 4.0 GHz.

Figs.8.8(a) and (b) illustrate the radiation patterns calculated using the regions D and R, respectively. A comparison of Figs.8.8(a) and (b) indicates that the intensity of the radiation field in Fig.8.8(b) is about 8 dB lower than that in Fig.8.8(a). Thus, the radiation from the region D constitutes the major component of the total radiation field.

In Fig.8.8(a), the E_ϕ component shows a symmetrical pattern with respect to the Z-axis. Although the symmetry of the E_θ component is somewhat distorted, the direction of the maximum remains near the Z-axis. On the other hand, in Fig.8.8(b), both E_ϕ and E_θ components are nonsymmetrical with respect to the Z-axis, and the direction of the maximum radiation deviates from the Z-axis.

Fig.8.8(c) shows the total radiation pattern, which is calculated taking account of the phase of the current of each region. It is found that the symmetry of the radiation pattern with respect to the Z-axis is slightly distorted because of an unfavourable phase condition between the two contributions of Figs.8.8(a) and (b). This distortion in the radiation pattern lowers the power gain on the Z-axis, causing the axial ratio to deteriorate. The power gain of 7.4 dB in Fig.8.8(a) is reduced to 6.9 dB in Fig.8.8(c). The axial ratio of 0.1 dB in Fig.8.8(a) deteriorates to 0.4 dB in Fig.8.8(c). It therefore follows that the region R is a basically undesirable region for the radiation.

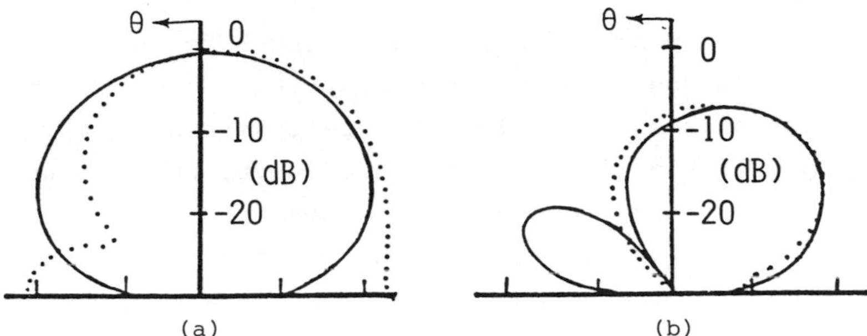

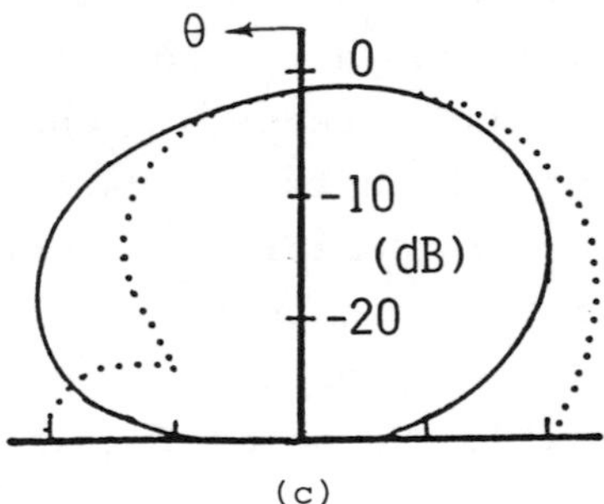

Fig.8.8 Effect of regions D and R on the radiation
(4.0 GHz, $\phi = 0°$ plane).
(a) radiation from decaying region D
(b) radiation from residual region R
(c) total radiation
——— E_ϕ
············ E_θ

8.7. REQUIREMENT FOR THE ARM LENGTH [4]

The facts mentioned above indicate that the antenna needs to have an arm of sufficient length to establish the region D. Although the current distribution of the region D is established at a low frequency, resulting in the radiation of a circularly polarised wave, the antenna arm length is not enough to obtain the wideband characteristics. It should be noted that, as the frequency is increased from the low frequency, a standing-wave current distribution appears with consequent deterioration in the radiation characteristics. This phenomenon is observed in transition frequencies from 1.8 GHz to 2.5 GHz, as shown in Fig.8.4.

At frequencies higher than 3.0 GHz, there exists a second minimum point, which is indicated by A_2 in Fig.8.4, in addition to the first minimum point. Once the second minimum is established, the current standing wave reduces and the radiation characteristics improve. It follows that usage over a wide range of frequencies requires the arm length to include the region up to the second minimum at the lowest operating frequency. In Fig.8.4 the second minimum in the current distribution is located at a circumference of about 2.6 λ.

8.8. SHORT CONICAL HELIX ANTENNA [5]

In the previous sections we have found that the current distribution of a monofilar conical helix antenna with a low pitch angle has two distinct regions, and that a decaying current region from the feed point to the first minimum point, called a region D, is responsible for the radiation pattern. When the extreme wideband characteristics are not required, a conical helix antenna may be constructed by a short antenna arm that supports only the region D in the current distribution. This type of antenna is designated as a short conical helix antenna (SCHA). In this section we consider the radiation characteristics of the SCHA.

The helix mounted on an infinite ground plane has the same configuration as used in the aforementioned study except for the total arm length. The total arm length L is 17.5 cm, corresponding to 2.33 λ at 4.00 GHz. It should be noted that the present antenna is made by cutting a long conical helix antenna at the first minimum point in the current distribution at 4.00 GHz (see Fig.8.3(c) or an inset of Fig.8.9(a)). By cutting the conical helix at the minimum point, the present antenna becomes a low silhouette of about 0.3 λ at 4.00 GHz.

The theoretical analysis is made at frequencies in the vicinity of 4.00 GHz. Fig.8.9 shows the theoretically determined current distributions and their phase progressions at 3.75 GHz, 4.25 GHz, and 5.00 GHz. At 4.25 GHz the current smoothly decays from the feed point with phase progression close to that in free space, whereas a standing wave appreciably appears near the arm end, as shown in Figs.8.9(a) and (c), as the frequency deviates from 4.25 GHz.

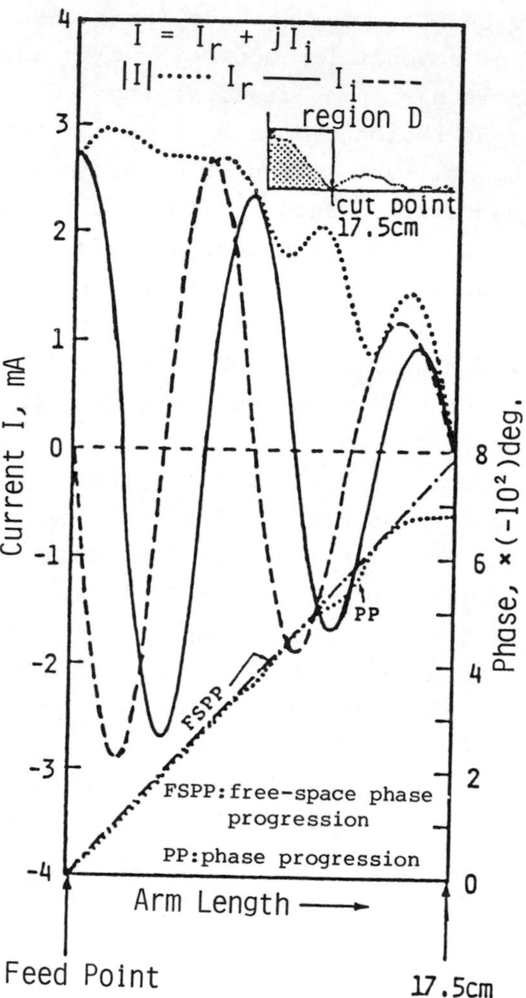

Fig.8.9(a) Current distribution at 3.75 GHz.

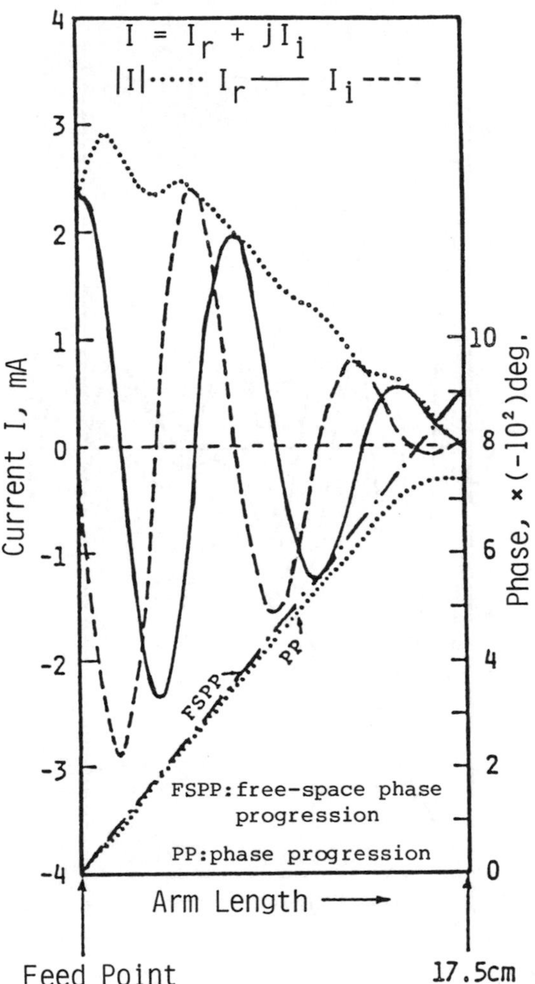

Fig.8.9(b) Current distribution at 4.25 GHz.

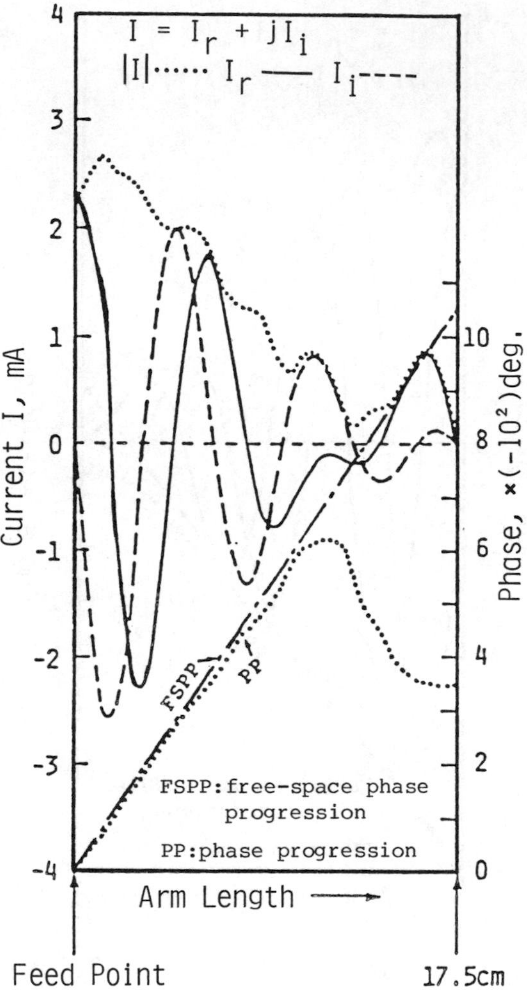

Fig.8.9(c) Current distribution at 5.00 GHz.

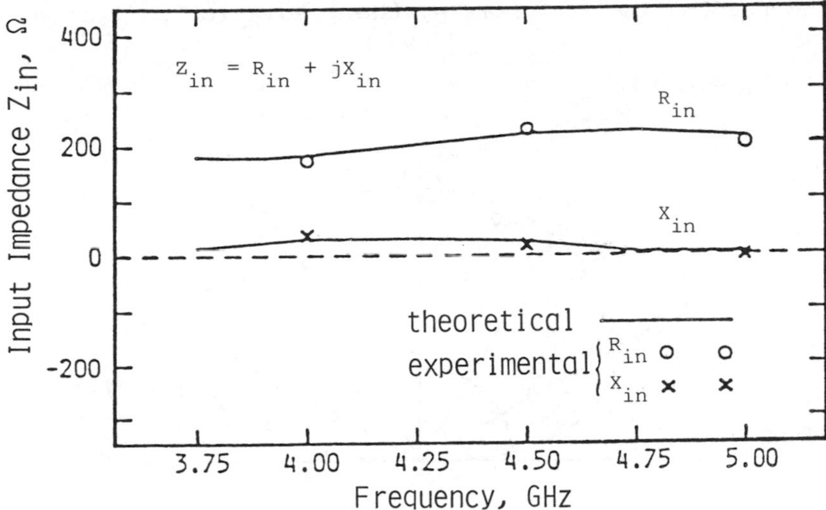

Fig.8.10 Input impedance.

Fig.8.10 illustrates the input impedance versus frequency. The theoretical input impedance indicates almost a pure resistance of the order of 200 ohms over a frequency range from 3.75 GHz to 5.00 GHz, showing agreement with experimental results. It should be noted that the experimental work is made by using the balanced type helix as shown in Fig.8.1(c) instead of using an infinite ground plane, and that the balanced type helix is fed by a bazooka balun designed for each frequency. The input impedance is measured on the basis of a conventional standing wave method, and the half value of the measured input impedance is plotted because of the conversion of the balanced type of Fig.8.1(c) to the real type of Fig.8.1(a).

A typical radiation pattern at 4.00 GHz is shown in Fig.8.11. Although the configuration is not symmetrical with respect to the Z-axis, the radiation pattern shows

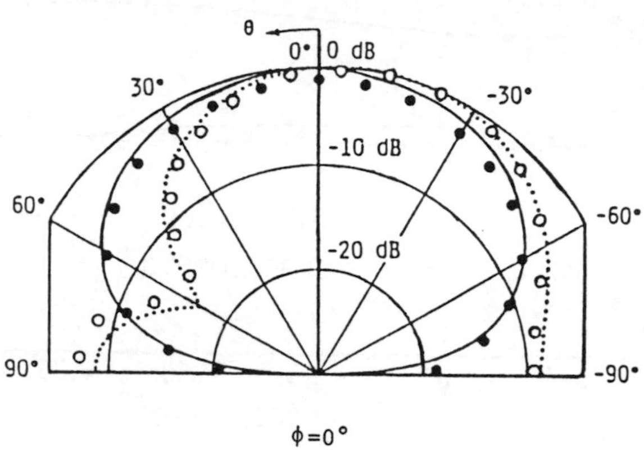

$\phi=0°$

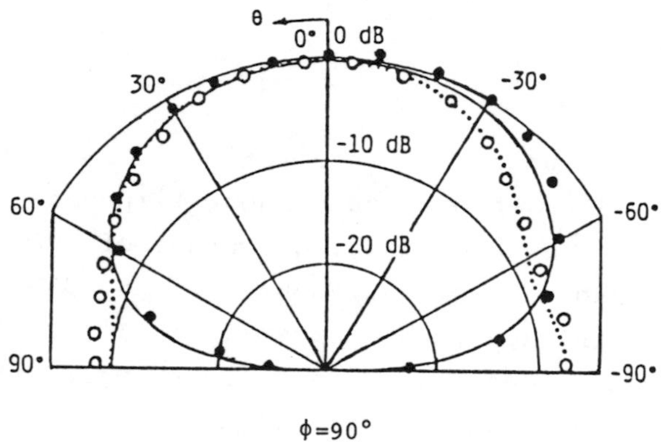

$\phi=90°$

Fig.8.11 Radiation pattern at 4.00 GHz.

Theoretical	Experimental
———— E_ϕ	● ● ● ● ● E_ϕ
-------- E_θ	○ ○ ○ ○ ○ E_θ

fairly good symmetry with respect to the Z-axis, with an average half-power beamwidth of about ± 40°. Excellent agreement is seen to exist between the theoretical and experimental results.

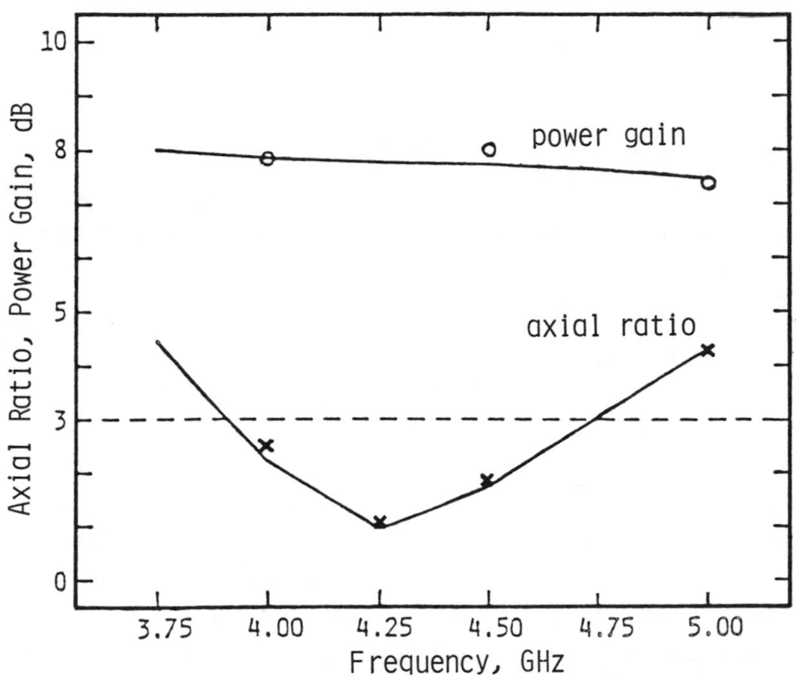

Fig.8.12 Axial ratio and power gain.

Theoretical ————

Experimental { power gain O
axial ratio ×

The frequency characteristics of the axial ratio and the power gain on the Z-axis are shown in Fig.8.12. It is found that, as the frequency is increased or decreased from 4.25 GHz, the axial ratio gradually deteriorates. The deterioration is due to the appearance of a standing wave in the current distribution, as shown in Figs.8.9(a) and (c). An axial ratio of less than 3.0 dB is obtained over a frequency range from 3.90 GHz to 4.75 GHz (1:1.2), where the power gain indicates about 7.7 dB. The experimental data are also presented in Fig.8.12.

8.9. CONCLUSION

The monofilar conical helix with a low pitch angle has been numerically analysed over a frequency range of 1.5 GHz to 6.0 GHz. The radiation pattern, axial ratio, input impedance and power gain are presented. It is revealed that the current distribution has two distinct regions, and that a decaying current from the feed point to a point of the first minimum is responsible for the radiation pattern. It is also found that an antenna for wideband usage should have an arm length for the current distribution which includes the second minimum point at the lowest frequency. The second minimum point is located at a circumference of about 2.6 λ in the present model.

A short conical helix antenna which has only a decaying current distribution has also been theoretically and experimentally investigated over a frequency range from 3.75 GHz to 5.00 GHz. The present antenna has a low silhouette of about 0.3 λ at 4.00 GHz. It is revealed that the short conical helix antenna radiates a circularly polarised wave with an axial ratio of less than 3 dB over a frequency range ratio of about 1:1.2, with a power gain of about 7.7 dB.

REFERENCES TO CHAPTER 8

[1] Chatterjee, J.S., "Radiation characteristics of a conical helix of low pitch angle", J. Appl. Phys., Vol. 26, 1955, pp.331-335.

[2] Yang, L.B. and Iizuka, K., "Experimental investigation on a method of lowering the silhouette of conical log-spiral antenna", IEEE Trans., AP-31, 1983, pp.347-352.

[3] Dyson, J.D., "The characteristics and design of the conical log-spiral antenna", IEEE Trans., AP-13, 1965, pp.488-499.

[4] Nakano, H., Mikawa, T. and Yamauchi, J., "Mono-filar helix antenna with low pitch angle", Proc. IEE, Vol. 131, 1984, pp.379-382.

[5] Nakano, H., Mikawa, T. and Yamauchi, J., "Investigation of a short conical helix antenna", IEEE Trans., AP-33, 1985, pp.1157-1160.

CHAPTER 9
Wire Scatterer for a Circularly Polarised Wave*

9.1. INTRODUCTION

The final chapter tackles scattering problems associated with applications of electromagnetic waves of circular polarisation.

In Section 9.2 we evaluate the radiation characteristics of a helix which is operated as a parasite, and show that the parasitic helix acts as a director for a circularly polarised wave. In addition, the application of a parasitic helix to a circularly polarised antenna is proposed and its performance is discussed.

In Section 9.3 we analyse a crossed-wire scatterer in order to use it as a reflector for a radar system using a circularly polarised wave (CPW-radar). In the CPW-radar a single antenna with a circular polariser is used concurrently as a transmitting antenna and a receiving antenna. Therefore, a reflector for the CPW-radar must generate a reflected wave of circular polarisation with the same rotational sense as that of an incident wave of circular polarisation. This reflected wave is called "double reflection". If a metallic planar plate is used as a reflector, the back-scattering wave from the reflector is not detected in the CPW-radar, because the back-scattering wave has the rotational sense opposite to that of the incident wave. This situation is often

* From papers (p-15),(p-25),and (c-13) listed at the end of this monograph.

referred to as "single reflection". If a crossed-wire scatterer has a junction point, it acts merely as a reflector of the single reflection for a circularly polarised wave. The problem to be treated in Section 9.3 is different from those of [1]-[7] in that a crossed-wire scatterer does not possess a junction point.

9.2. SCATTERING FROM PARASITIC HELIX [8]

9.2.1. Parasitic helix illuminated by a circularly polarised plane wave

This section deals with the scattering characteristics of a parasitic helix illuminated by a circularly polarised plane wave (CPPW) with unit amplitude ($\bar{E}=[\hat{x}-j\hat{y}]e^{-j\beta z}$). Fig.9.1 shows the configuration and the co-ordinate system of a parasitic helix with a tapered section. The parasitic helix has the geometrical parameters which are common in a conventional axial-mode helical antenna fed

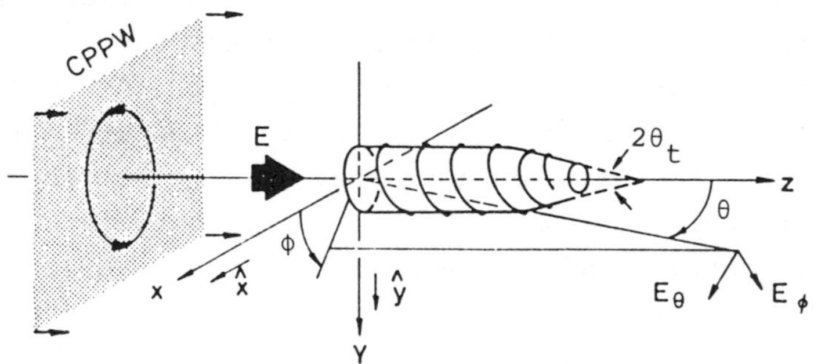

Fig.9.1 Configuration and co-ordinate system of parasitic helix with tapered section. $\hat{x}$ and $\hat{y}$ are unit vectors

from a coaxial line [see Chapter 6]. In the present model, the circumference of the helical cylinder c is 10 cm (=1λ at 3.0 GHz), the pitch angle α is 12.5°, and the number of helical turns t is 6. The tapering-cone angle θ_t is taken to be the same as the pitch angle.

The testing frequency of this analysis is chosen to be a range of 2.3 GHz to 4.0 GHz. This is based on the fact that the conventional helical antenna operates in the axial mode over a circumference range of (3/4)λ to (4/3)λ.

Figs.9.2(a) and (b) show the current distributions at 2.3 GHz and 4.0 GHz, respectively, which are determined by a modified form of eqn.(1.63), or eqn.(1.64). The far-field patterns of the scattered fields are also presented.

At 2.3 GHz the parasitic helix is about to establish a travelling current with constant amplitude. The scattered field tends to form a unidirectional pattern, and the axial ratio on the +Z-axis is 3.8 dB. The numerical result at 4.0 GHz indicates that a travelling current with nearly constant amplitude is induced along the helical wire, i.e., the surface wave is excited. This fact holds true over a frequency range of 2.6 GHz to 4.0 GHz.

Using the obtained current-phase progression, we calculate the relative phase velocity p, which is defined as the phase velocity relative to the velocity of light in free space. Fig.9.3 shows the calculated relative phase velocity as a function of frequency. It is found that the current approximately satisfies the in-phase condition of the end-fire radiation in spite of the change in frequency. This leads to wideband characteristics for the scattered field, which exhibits a unidirectional pattern with a maximum in the direction of the +Z-axis. Further calculation concerning the scattered field shows that the axial ratio is in a range of 0.6 dB to 2.1 dB over a frequency range of 2.6 GHz to 4.0 GHz.

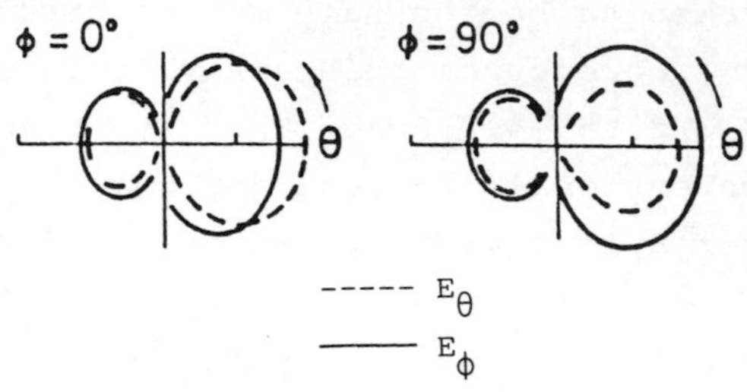

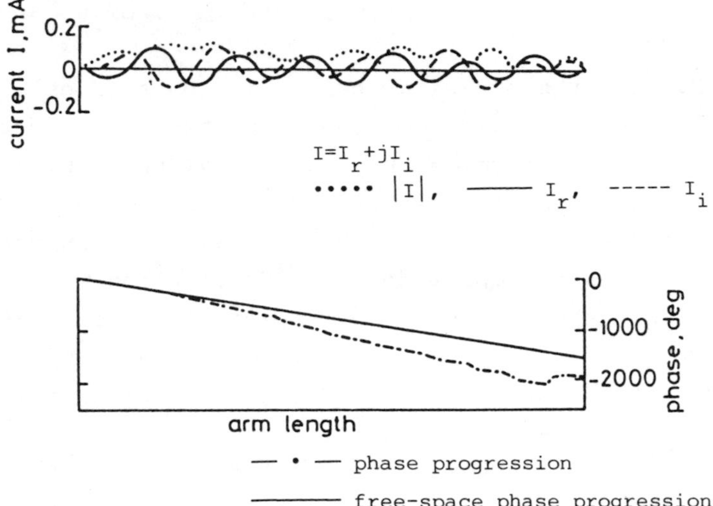

(a) 2.3 GHz

Fig.9.2(a) Current distribution and far-field patterns of scattered fields when a tapered parasitic helix is illuminated by a circularly polarised plane wave.

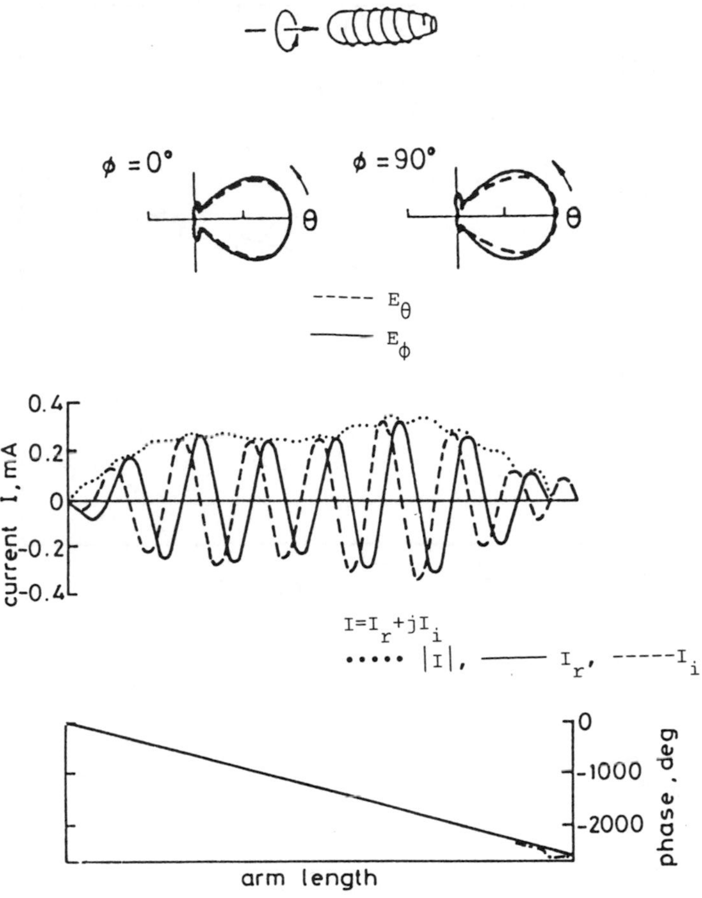

(b) 4.0 GHz

Fig.9.2(b) Current distribution and far-field patterns of scattered fields when a tapered parasitic helix is illuminated by a circularly polarised plane wave.

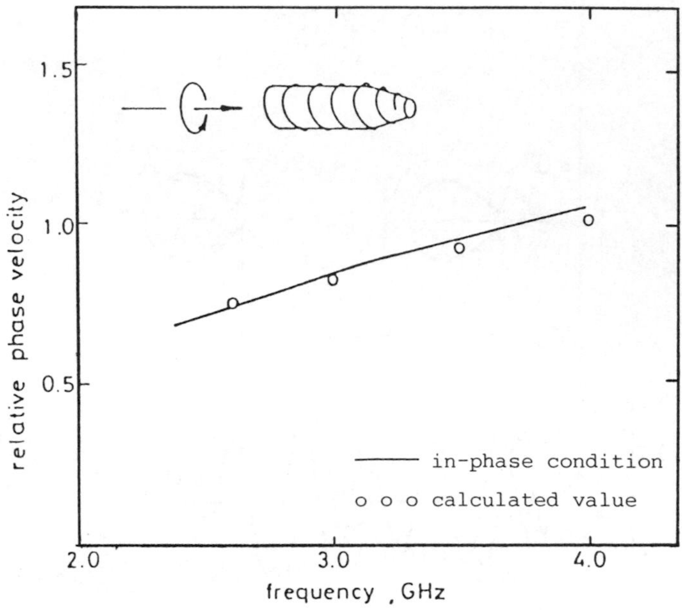

Fig.9.3 Relative phase velocity as a function of frequency.

9.2.2. Application of parasitic helix to circularly polarised antenna

We consider an antenna system consisting of a circularly polarised antenna and a parasitic helix. To excite a parasitic helix, a square spiral antenna is chosen as an exciter, as shown in Fig.9.4. The present spiral antenna as a single radiating element in free space has an axial ratio of 1.2 dB, a power gain of 3.8 dB, and an input impedance of nearly a pure resistance of 240 ohms at a testing frequency of 3.0 GHz.

First, consideration is given to the basic characteristics of the antenna system with ϕ_r=0°, where ϕ_r

is the rotation angle, or the initial winding angle of the helical wire, as defined in Fig.9.4. The spacing d between the spiral and the parasitic helix is taken to be 0.5λ.

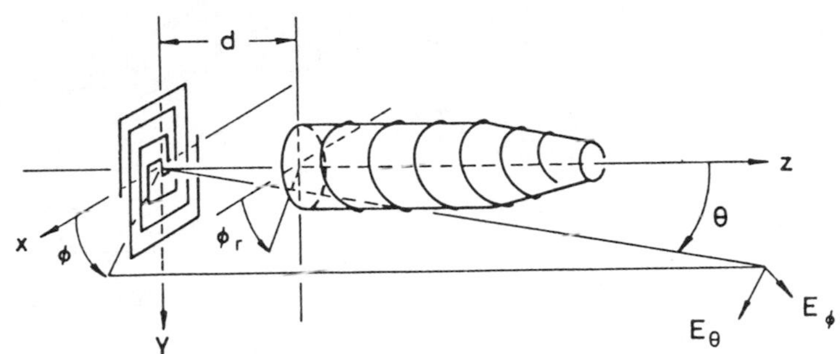

Fig.9.4 Configuration of antenna system composed of square spiral antenna and tapered parasitic helix.

Fig.9.5(a) shows the radiation pattern (power pattern) of the antenna system. For comparison, the radiation pattern of the isolated spiral antenna is shown in Fig.9.5(b). The average half-power beamwidth of the E_θ- and E_ϕ-components is changed from ±49° to ±26°, and the power gain is increased from 3.8 dB to 7.0 dB, with the help of the parasitic helix. The antenna system exhibits an axial ratio of 0.6 dB.

Further analysis shows that the current distribution of the antenna system remains basically the same even when the parasitic helix is rotated. This leads to the fact that, in spite of the rotation of the parasitic helix, the directive properties of the parasitic helix are maintained, and the fact that the radiation pattern and

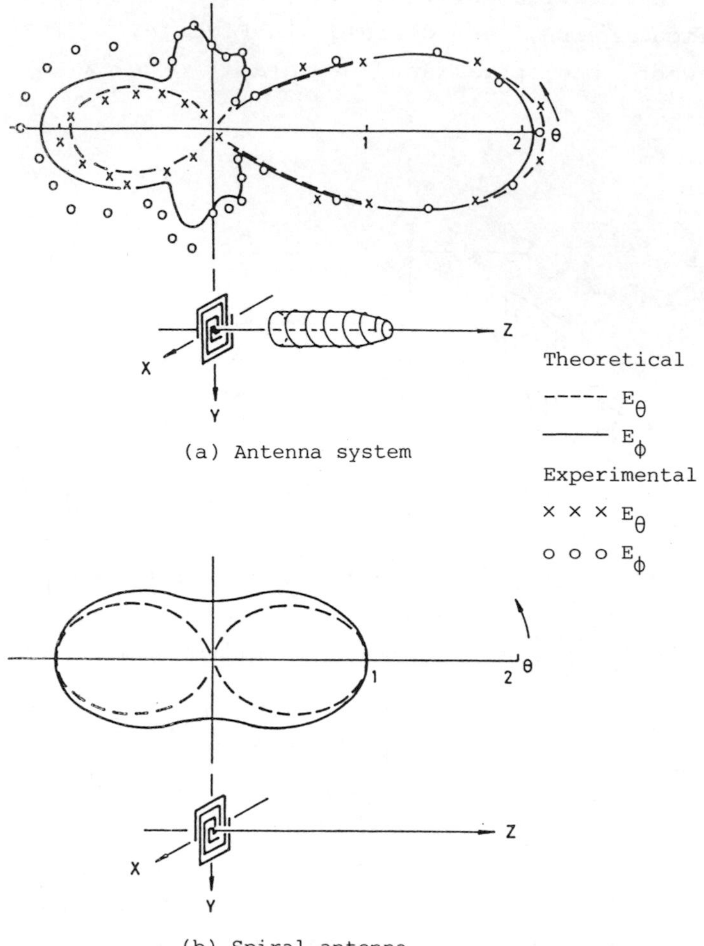

(a) Antenna system

(b) Spiral antenna

Fig.9.5 Radiation patterns.
The radiation pattern of the antenna system is normalised to a maximum value of the field components of the isolated spiral antenna. The input power to the spiral in (a) is the same as that in (b).
c=10 cm (=1λ), t=6 turns, 3.0 GHz
ϕ=0° plane, ϕ_r=0°, d=5.0 cm (=0.5λ)

the input impedance of the antenna system remain almost unchanged.

Secondly, consideration is given to the behaviour of the antenna system when the spacing d between the spiral and the parasitic helix is changed under the condition of $\phi_r=0°$. The behaviour to be expected is that as the spacing d is increased the current amplitude induced on the parasitic helix decreases, with a subsequent decrease in the radiation from the parasitic helix. Fig.9.6 shows the relation between the power gain and the spacing d, together with the axial ratio. The gain shows the highest value of 7.0 dB when the spacing d is between 0.4λ and 0.5λ, and a tendency to decrease as the spacing d is increased from 0.5λ to 2.0λ. The present arrangement, therefore, has an advantage of being able to control the gain by only adjusting the spacing d (for a conventional helix, the control of the gain must be made by the change in the diameter of the helical cylinder or in the number of turns).

It is also observed that in Fig.9.6 the gain decreases as the spacing d becomes less than 0.4λ. According to the calculation, the current on the parasitic helix is disturbed owing to the mutual coupling. The standing wave along the arm becomes appreciable, and the change in phase gradually loses its linearity. Consequently, the radiation in the direction of the +Z-axis from the parasitic helix is decreased, and thus the gain of the antenna system is decreased.

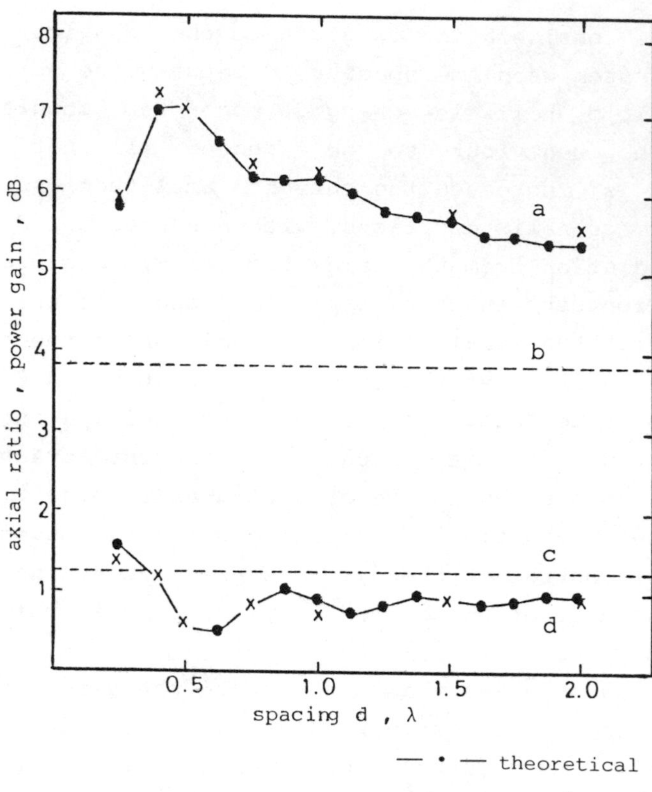

— • — theoretical

× × × × experimantal

a Power gain of antenna system

b Power gain of isolated spiral

c Axial ratio of isolated spiral

d Axial ratio of antenna system

Fig.9.6 Power gain and axial ratio as a function of spacing d. c=10 cm (=1λ), t=6 turns, 3.0 GHz

9.3. A CROSSED-WIRE SCATTERER FOR AN INCIDENT WAVE OF CIRCULAR POLARISATION [9]

The purpose of this section is to explore the possibility that a crossed-wire scatterer without a junction point acts as a reflector whose performance corresponds to a dihedral corner reflector [10], in which the back-scattering wave has the same rotational sense as that of an incident wave of circular polarisation by virtue of double reflection.

9.3.1. A single straight crossed-wire scatterer

Attention is focused on the back-scattering wave from a single straight crossed-wire scatterer to obtain basic data. The configuration of the scatterer is shown in Fig.9.7. Both the horizontal and the vertical wires are half-wavelength in length, i.e., $2L_1 = 2L_2 = \lambda/2$, where λ is the wavelength of an incident wave. In the following analysis a wavelength of λ=0.032 [m] (a frequency of 9.375GHz) is used. Each wire is perfectly conducting, and is thin relative to its length as well as to the wavelength of the incident wave (the radius of wire a_0 is 0.00469 λ). The spacing between the horizontal and the vertical wires is designated S.

The axial ratio of the back-scattering wave (on the Z-axis) is shown in Fig.9.8 as a function of the spacing S. The scatterer is illuminated by a plane wave of circular polarisation of left handed sense from the Z-axis direction or by an electric field of $\overline{E}^{in} = E_0(\hat{\theta}-j\hat{\phi})\exp(j\beta z)$ with an angle of incidence $\theta=0°$, where $\hat{\theta}$ and $\hat{\phi}$ are unit vectors of spherical co-ordinates. Except for

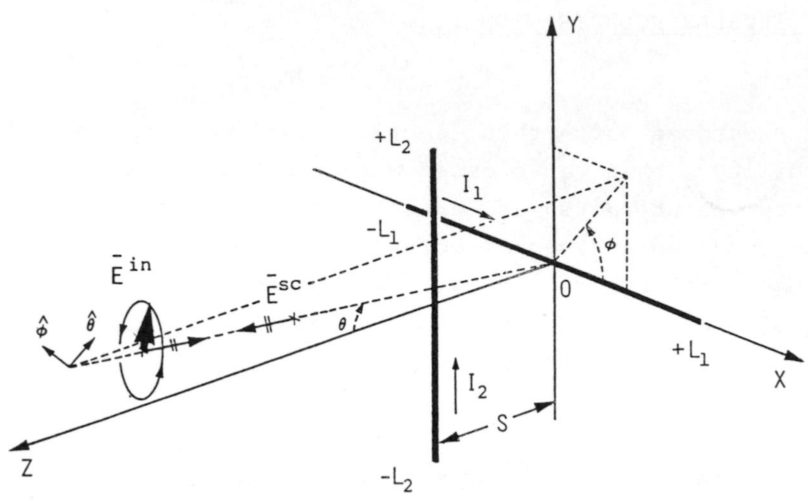

Fig.9.7 Configuration of crossed-wire scatterer.

$S = 0.125n\lambda$ ($n=0,1,2,...$), the back-scattering wave is elliptically polarised. In Fig.9.8 the rotational sense of the back-scattering wave is expressed by two curves: the broken curve is for the right handed sense; the solid curve is for the left handed sense. We should recall that the senses are determined by a relation between the directions of the rotation of electric field and the wave propagation. With either the right handed sense or the left handed sense, an axial ratio within 3 dB is repeated over ranges of $S=0.25n\lambda \pm 0.03\lambda$ ($n=1,2,...$).

The back-scattering wave $\bar{E}^{sc}$ ($= E_\theta \hat{\theta} + E_\phi \hat{\phi}$) is decomposed into two components which are circularly polarised with different intensities and rotational senses (left handed sense and right handed sense),

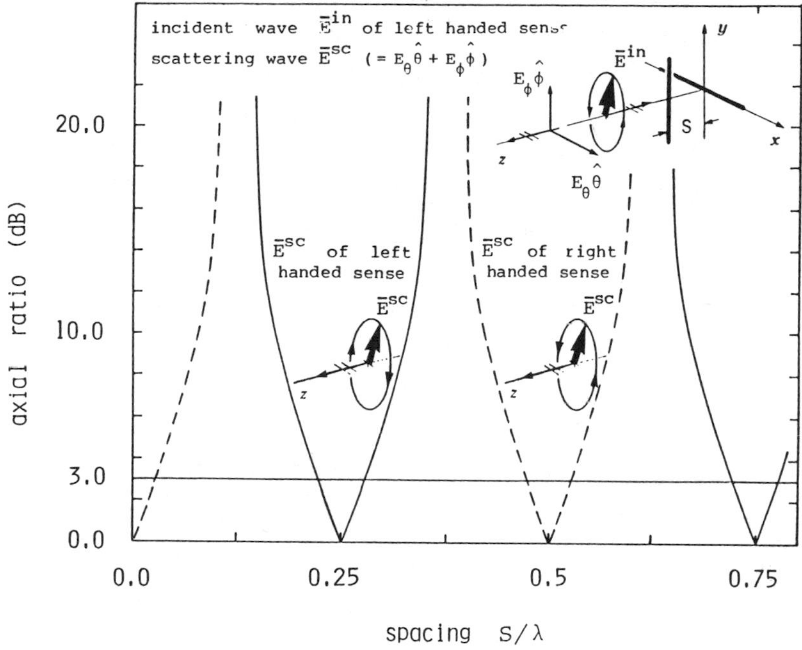

Fig.9.8 Axial ratio observed in the normal direction as a function of spacing S.
right handed sense - - - - - -
left handed sense ————

$$\bar{E}^{sc} = E_L(\hat{\theta} + j \hat{\phi}) + E_R(\hat{\theta} - j \hat{\phi}). \quad (9.1)$$

Using a relation of $|E_\theta|^2 + |E_\phi|^2 = 2\times(|E_L|^2 + |E_R|^2)$, the back-scattering cross-section (BSCS) σ is defined as

$$\sigma = \sigma_L + \sigma_R , \quad (9.2)$$

$$\text{where} \quad \sigma_{i=L,R} = \lim_{r\to\infty} \frac{4\pi r^2 |E_i|^2}{|E_0|^2} \quad (9.3)$$

in which r is the distance from the scatterer to an observation point.

A back-scattering wave which has the same rotational sense as an incident wave of circular polarisation is called a "fundamental component", while a back-scattering wave whose rotational sense is opposite to that of the incident wave is called a "cross component". In the present consideration the BSCS for the fundamental component is given by σ_L, since the incident wave is a circularly polarised wave of left handed sense.

Fig.9.9 shows the BSCS's as a function of the angle of incidence θ. The difference in the BSCS between the fundamental and the cross components is 6 dB at the half-BSCS angle (angle at which the BSCS for the

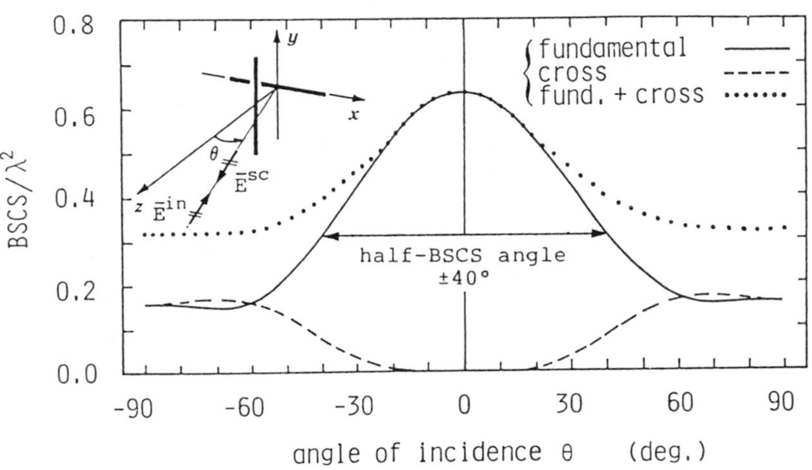

Fig.9.9 Back-scattering cross sections as a function of angle of incidence θ : ϕ=0° plane (XZ-plane); spacing S=0.25λ.

fundamental component is 3 dB down from the maximum value). For an angle of incidence $\theta = \pm 90°$, the current is induced only on the vertical wire, and the back-scattering wave is linearly polarised. This means that the BSCS for the fundamental component is half of the total BSCS, or the cross component appears with the same intensity. Since the appearance of the cross component deteriorates the transformation efficiency of the incident energy to the fundamental component in the back-scattering wave, it is desirable for the cross component to be as small as possible.

9.3.2. Bent crossed-wire scatterer and its arrays

In this section a bent crossed-wire scatterer is proposed to improve the radiation performance of the straight crossed-wire scatterer.

Fig.9.10 shows the configuration of a single crossed-wire scatterer consisting of bent arms. The arm length is taken to be $2L_1=2L_2=0.5\lambda$, with a bend length of $L=0.125\lambda$. The spacing S is chosen to be 0.25λ so as to produce a back-scattering wave whose rotational sense is the same as that of an incident wave of circular polarisation.

The idea of bending the arms comes from the fact found in a half-wavelength dipole antenna. The current amplitude of a half-wavelength dipole antenna bent at a middle point ($L=0.125\lambda$) of each arm indicates a maximum value when a bend angle is about 90° [11]. The idea is applicable to a wire scatterer as a counterpart; when a half-wavelength isolated wire scatterer is illuminated by a linearly polarised wave, the current amplitude of the

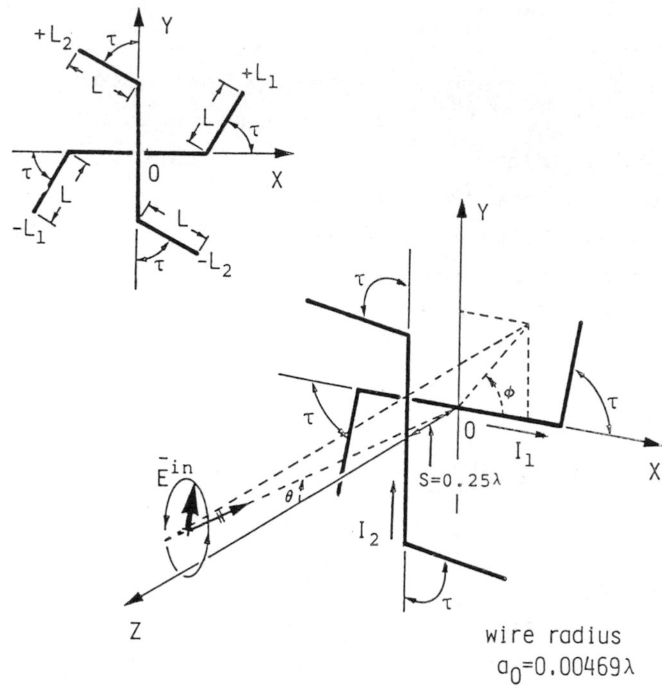

Fig.9.10 Configuration of bent crossed-wire scatterer.

wire scatterer shows a maximum value at a bend angle τ of about 90°, resulting in a maximum value of the BSCS. The described behaviour of the current on the isolated wire scatterer also holds for an incident wave of circular polarisation. It is noted that, for the bent crossed-wire scatterer consisting of two arms as shown in Fig.9.10, the bend angle should be determined so that the BSCS may give a maximum value, taking the mutual coupling between the arms into account.

Fig.9.11 shows the BSCS for the fundamental component and the axial ratio as a function of bend angle τ. The calculations are carried out under the condition that an incident wave of circular polarisation of left handed sense illuminates the bent crossed-wire scatterer from an angle of incidence $\theta=0°$. The excellent axial ratio leads to an extremely small BSCS for the cross component with consequent disappearance in Fig.9.11. It is found that the BSCS for the fundamental component exhibits a maximum value of $0.826\lambda^2$ at a bend angle of about 85° with almost entirely circular polarisation, and that the maximum value increases about 30 % above the BSCS for $\tau=0°$ (straight-arm crossed-wire scatterer). The favourable

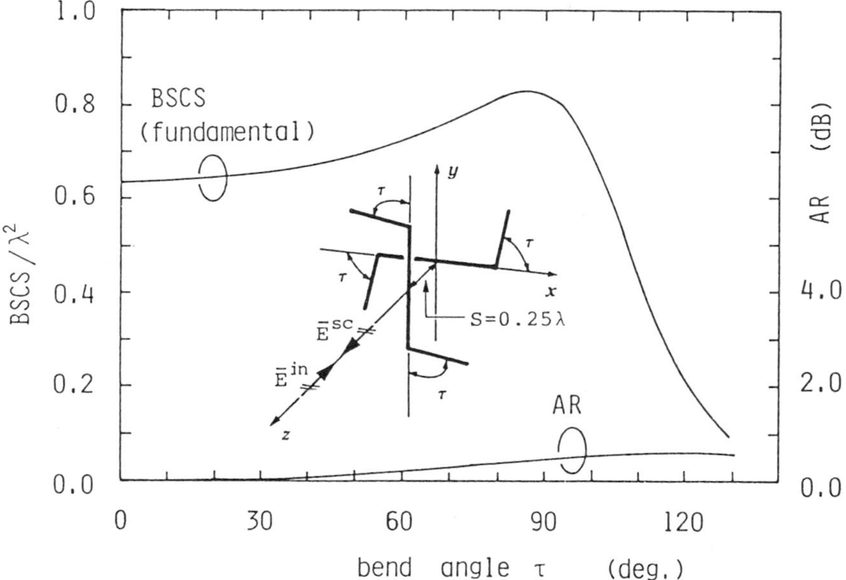

Fig.9.11 Back-scattering cross section for fundamental component and axial ratio as a function of bend angle τ.

increase in the BSCS results from the enhancement in the current amplitudes on the arms. Although increasing the current amplitudes can be realised by an alternative way of truncating the wires slightly, bending the wires has an advantage in that an aperture area for $\tau=85°$ is only 30 % of that for $\tau=0°$.

Fig.9.12 shows the BSCS's of the bent crossed-wire scatterer with $\tau=85°$ as a function of the angle of incidence θ. Comparing Fig.9.12 with Fig.9.9, we observe that the BSCS for the fundamental component of the bent crossed-wire scatterer increases near the Z-axis, while the BSCS for the cross component remains almost the same as that of the straight crossed-wire scatterer. Hence, a relative intensity of the fundamental component to an

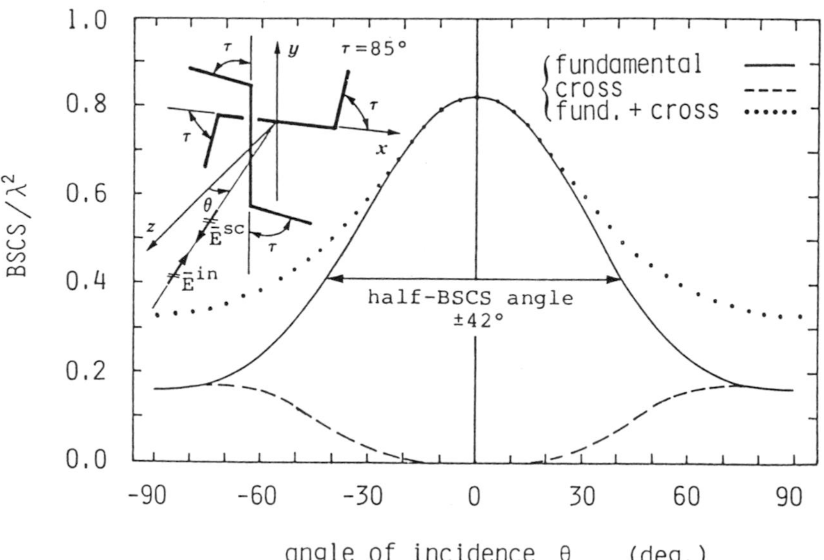

Fig.9.12 Back-scattering cross section as a function of angle of incidence θ: $\phi=0°$ plane; spacing $S=0.25\lambda$.

intensity of the cross component increases near the Z-axis. The difference in the BSCS between the fundamental and the cross components at the half-BSCS angle improves to 7 dB from 6 dB for the straight crossed-wire scatterer. Further improvement will be found in an array consisting of the bent crossed-wire scatterers.

On the basis of the results mentioned above, an array is constructed using 3×3 bent crossed-wire scatterers with a bend angle of $\tau=85°$. Fig.9.13 shows the current amplitudes of the array elements separated at a distance of $D=0.745\lambda$, where the BSCS for the fundamental component becomes a maximum value of $213\lambda^2$. The maximum BSCS of the array is about 258 times as large as that of the single bent crossed-wire scatterer shown in Fig.9.11. The average of the current amplitudes of the array is about 1.8 times as large as that of the single bent crossed-wire scatterer, due to the mutual effects among the array elements. An approximate calculation of $(1.8\times9)^2$ for nine array elements accounts for a multiplicative factor of 258. It is also confirmed that, as the distance D between the array elements is increased, a maximum value of the BSCS for the fundamental component approaches a value which is 81 [$= (3\times3)^2$] times as large as that of the single bent crossed-wire scatterer.

A comparison between the array and a dihedral corner reflector shows that the maximum BSCS of the former is about 1.8 times as large as that of the latter, both having the same aperture area. A comparison between them also indicates that the array structure has an advantage in that the spacing S is independent of an increase of the aperture area, i.e., $S=0.25\lambda$. On the other hand, the dihedral corner reflector is forced to occupy a deeper spacing between the aperture plane and the intersection of the two metallic planes, increasing the aperture area.

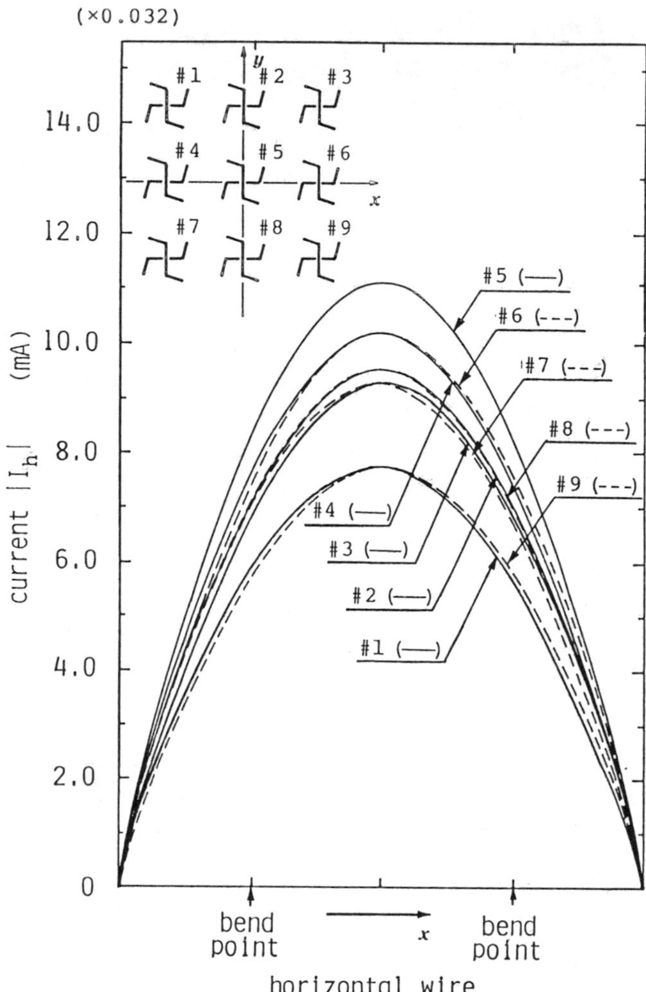

Fig.9.13(a) Current amplitudes on 3×3 bent crossed-wire scatterers (horizontal wires) : bend angle τ=85°; spacing S=0.25λ; distance between elements D=0.745λ; E_0=1 V/m.

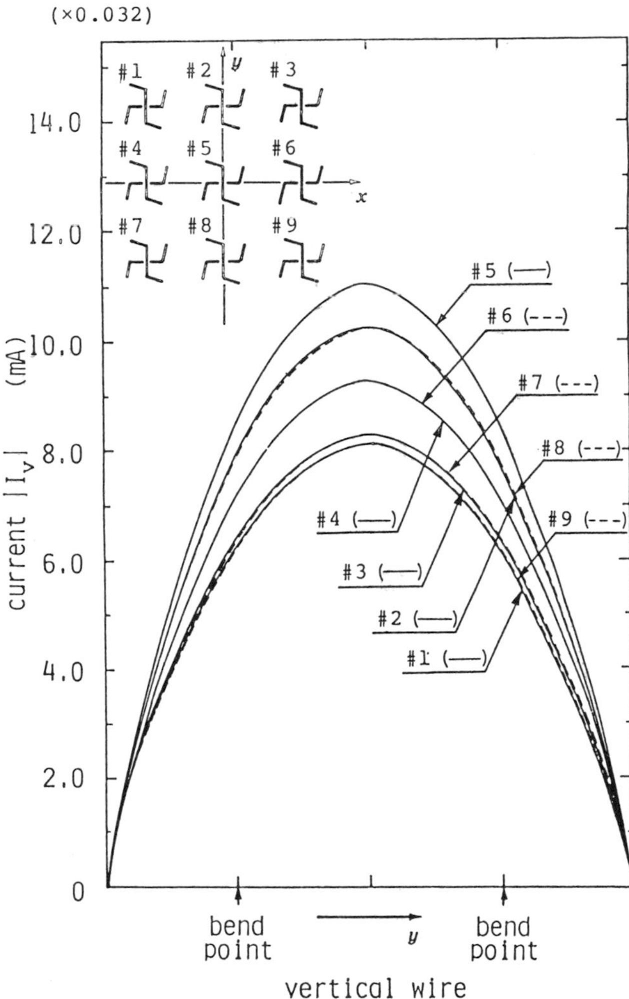

Fig.9.13(b) Current amplitudes on 3×3 bent crossed-wire scatterers (vertical wires) : bend angle τ=85°; spacing S=0.25λ; distance between elements D=0.745λ; E_0=1 V/m.

Fig.9.14 shows the BSCS as a function of the angle of incidence θ. The half-BSCS angle is ± 6.5°, over which the BSCS for the cross component is less than -31 dB and is not appreciable in Fig.9.14.

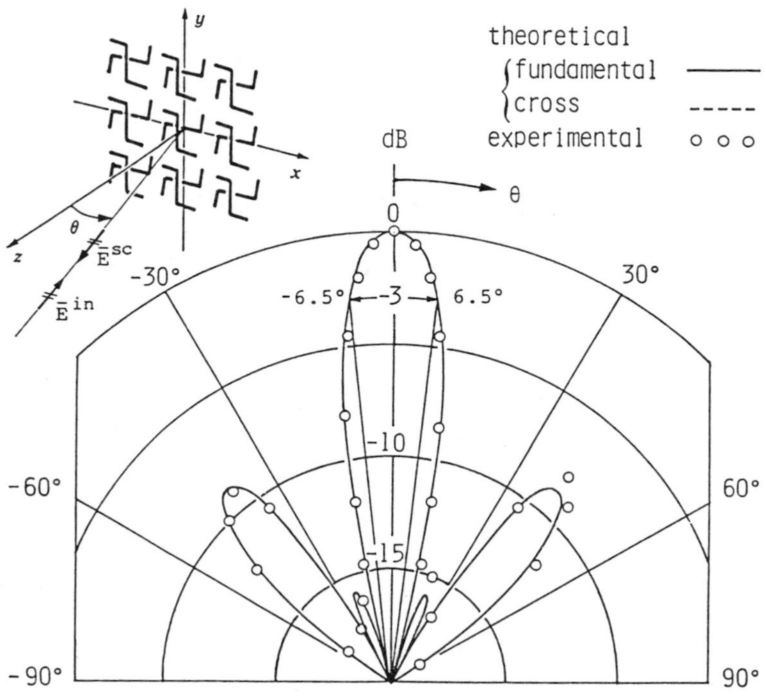

Fig.9.14 Back-scattering cross sections of array of 3×3 bent cross-wire scatterers : ϕ=0° plane; bend angle τ=85°; spacing S=0.25λ; distance between elements D=0.745λ.

REFERENCES TO CHAPTER 9

[1] Taylor, C.D., Lin, S.M. and McAdams, H.V., "Scattering from crossed wires", IEEE Trans., AP-18, 1970, pp.133-136.

[2] Chao, H.H. and Strait, B.J., "Radiation and scattering by configurations of bent wires with junctions", IEEE Trans., AP-19, 1971, pp.701-702.

[3] Butler, C.M., "Currents induced on a pair of skew crossed wires", IEEE Trans., AP-20, 1972, pp.731-736.

[4] King, R.W.P., "Currents induced in a wire cross by a plane wave incident at an angle", IEEE Trans., AP-25, 1977, pp.775-781.

[5] King, R.W.P. and Sander, B.H., "Analysis of the currents induced in a general wire cross by a plane wave incident at an angle with arbitrary polarization", IEEE Trans., AP-29, 1981, pp.512-520.

[6] Agrawal, V.D. and Imbriale, W.A., "Design of a dichroic cassegrain subreflector", IEEE Trans., AP-27, 1979, pp.466-473.

[7] Pelton, E.L. and Munk, B.A., "Scattering from periodic arrays of crossed dipoles", IEEE Trans., AP-27, 1979, pp.323-330.

[8] Nakano, H., Yamane, T. and Yamauchi, J., "Directive properties of parasitic helix and its application to circularly polarised antenna", Proc. IEE, Vol.130, 1983, pp.391-396.

[9] Nakano, H., Yoshizawa, A., and Yamauchi, J., "Characteristics of a crossed-wire scatterer without a junction point for an incident wave of circular polarization"., IEEE Trans., AP-33, No.4, 1985, pp.409-415.

[10] Johnson, R.C. and Jasik, H., "Antenna engineering handbook, 2nd ed.," New York: McGraw-Hill, 1984, Ch.17-27.

[11] Nakano, H., Yamauchi, J., Kawashima, K., and Hirose, K., "Effects of arm bend and asymmetric feeding on dipole antennae", Int. J. Electron., Vol.55, 1983, pp.353-364.

Papers written by the author relevant to this monograph (I)

(p-1) Nakano, H. and Yamauchi, J., "The balanced helices radiating in the axial mode", Trans. IECE Japan, Vol.62-B, No.7, 1979, pp.707-709.

(p-2) Nakano, H., "The simplified expression for the kernel of Mei's integral equation", Trans. IECE Japan, Vol.62-B, No.11, 1979, pp.1058-1059.

(p-3) Nakano, H., Yamauchi, J and Koizumi, H., "The theoretical and experimental investigations of the two-wire square spiral antenna", Trans. IECE Japan, Vol.E63, No.5, 1980, pp.337-343.

(p-4) Nakano, H., Yamauchi, J., Mimaki, H. and Sugano, M., "The balanced helical antenna radiating right- and left-hand circularly-polarized-waves", Trans. IECE Japan, Vol.63-B, No.8, 1980, pp.743-750.

(P-5) Nakano, H. and Yamauchi, J., "Radiation characteristics of helix antenna with parasitic elements", Electronics Letters, Vol.16, No.18, 1980, pp.687-688.

(p-6) Yamauchi, J., Nakano, H. and Mimaki, H., "Balanced helical antenna with tapered open ends", Trans. IECE Japan, Vol.J64-B, No.4, 1981, pp.279-286.

(p-7) Yamauchi, J. and Nakano, H., "Axial ratio of balanced helical antenna and ellipticity measurement of incident wave", Electronics Letters, Vol.17, No.11, 1981, pp.365-366.

(p-8) Nakano, H., Yamauchi, J and Hashimoto, S., "Sunflower spiral antenna", Trans. IECE Japan, Vol.E64, No.12, 1981, pp.763-769.

(p-9) Nakano, H., Yamauchi, J. and Iio, S.,"Tapered backfire helical antenna with loaded termination", Electronics Letters, Vol.18, No.4, 1982, pp.158-159.

(p-10) Nakano, H., "The integral equations for a system composed of many arbitrarily bent wires", Trans. IECE Japan, Vol.E65, No.6, 1982, pp.303-309.

(p-11) Nakano, H. and Yamauchi, J., "Characteristics of modified spiral and helical antennas", IEE Proc., Vol.129, Pt.H, No.5, 1982, pp.232-237.

(p-12) Nakano, H., Yamauchi, J., Mimaki, H. and Iio, S., "Backfire bifilar helical antenna with tapered feed end", Int. J. Electronics, Vol.54, No.2, 1983, pp.279-286.

(p-13) Nakano, H., Yamauchi, J. and Hashimoto, S., "Numerical analysis of 4-arm Archimedean spiral antenna", Electronics Letters, Vol.19, No.3, 1983, pp.78-80.

(p-14) Nakano, H., Harrington, R.F., "Integral equations on electromagnetic coupling to a wire through an aperture", Trans. IECE Japan, Vol.E66, No.6, 1983, pp.383-389.

(p-15) Nakano, H., Yamane, T. and Yamauchi, J., "Directive properties of parasitic helix and its application to circularly polarised antenna", IEE Proc., Vol.130, Pt.H, No.6, 1983, pp.391-396.

(p-16) Nakano, H., Yamauchi, J. and Nogami, K., "Effects of wire radius and arm bend on a rectangular spiral antenna", Electronics Letters, Vol.19, No.23, 1983, pp.957-958.

(p-17) Nakano, H., Yamauchi, J. and Sugiyama, Y., "Parasitic effects and polarization diversity of Archimedean spiral antenna", Trans. IECE Japan, Vol.J66-B, No.11, 1983, pp.1418-1425.

(p-18) Nakano, H., Hirose, K. and Yamauchi, J., "Spiral antenna with two off-center sources", Trans. IECE Japan, Vol.E67, No.6, 1984, pp.309-314.

(p-19) Nakano, H., Iio, S. and Yamauchi, J., "Frequency characteristics of tapered backfire helical antenna with loaded termination", IEE Proc., Vol.131, Pt.H, No.3, 1984, pp.147-152.

(p-20) Nakano, H., Asaka, N. and Yamauchi, J., "Short helical antenna array fed from a waveguide", IEEE Trans., Vol.AP-32, No.8, 1984, pp.836-840.

(p-21) Nakano, H., Asaka, N. and Yamauchi, J., "Radiation characteristics of short helical antenna and its mutual coupling", Electronics Letters, Vol.20, No.5, 1984, pp.202-204.

(p-22) Nakano, H., Hirose, K. and Yamauchi, J., "Numerical analysis of asymmetrical spiral antenna", IEEE Trans., Vol.AP-33, No.6, 1985, pp.676-678.

(p-23) Nakano, H., Yamane, T. and Yamauchi, J., "Hallen-type integral equations for a system composed of wires and slots", Trans. IECE Japan, Vol.E67, No.9, 1984, pp.516-522.

(p-24) Nakano, H., Mikawa, T. and Yamauchi, J., "Monofilar conical-helix antenna with low pitch angle", IEE Proc., Vol.131, Pt.H, No.6, 1984, pp.379-382.

(p-25) Nakano, H., Yoshizawa, A. and Yamauchi, J., "Characteristics of a crossed-wire scatterer without a junction point for an incident wave of circular polarization", IEEE Trans., Vol.AP-33, No.4, 1985, pp.409-415.

(p-26) Nakano, H., Mikawa, T. and Yamauchi, J., "Investigation of a short conical helix antenna", IEEE Trans., Vol.AP-33, No.10, 1985, pp.1157-1160.

(p-27) Nakano, H., Nogami, K., Arai, S., Mimaki, H. and Yamauchi, J., "A spiral antenna backed by a conducting plane reflector", IEEE Trans., Vol.AP-34, No.6, 1986, pp.791-796.

(p-28) Nakano, H., Samada, Y. and Yamauchi, J., "Axial mode helical antennas", IEEE Trans., Vol.AP-34, No.9, 1986, pp.1143-1148.

(p-29) Nakano, H., Sato, M. and Yamauchi, J., "Generation of a wide circularly polarised wave from a conical-helix antenna", Electronics Letters, Vol.23, No.8, 1987, pp.387-388.

Conference papers written by the author relevant to this monograph (II)

(c-1) Nakano, H. and Yamauchi, J., "The two-wire square spiral antenna", Proceedings of IECE International conference on antennas and propagation, Sendai Japan, 1978, pp.137-140.

(c-2) Nakano, H. and Yamauchi, J., "A theoretical investigation of the two-wire round spiral antenna -Archimedean type-", Proc. IEEE International symposium on antennas and propagation, Seattle, 1979, pp.387-390.

(c-3) Nakano, H. and Yamauchi, J., "The balanced helices radiating in the axial mode", Proc. IEEE International symposium on antennas and propagation, Seattle, 1979, pp.404-407.

(c-4) Nakano, H. and Yamauchi, J., "Sunflower spiral antenna", Proc. IEEE International symposium on antennas and propagation, Quebec, 1980, pp.709-712.

(c-5) Nakano, H., Yamauchi, J. and Mimaki, H., "Tapered balanced helices radiating in the axial mode", Proc. IEEE International symposium on antennnas and propagation, Quebec, 1980, pp.700-703.

(c-6) Nakano, H., Yamauchi, J. and Mimaki, H., "The short balanced helix radiating in the axial mode", Proc. URSI International symposium on electromagnetic waves, Munich, 1980, pp.343A/1-343A/4.

(c-7) Nakano,H. and Yamauchi, J., "The radiation characteristics of two-wire square spiral antenna", Proc. URSI International symposium on electromagnetic waves, Munich, 1980, pp.341A/1-341A/4.

(c-8) Nakano, H. and Yamauchi, J., "Characteristics of modified spiral- and helical- antenna", Proc. IEE Second International symposium on antennas and propagation, York, 1981, pp.293-297.

(c-9) Nakano, H. and Yamauchi, J., "Axial ratios of spiral antennas", Proc. IEEE International symposium on antennas and propagation, Los Angeles, 1981, pp.679-682.

(c-10) Yamauchi, J., Nakano, H. and Mimaki, H., "Backfire bifilar helical antenna with tapered feed end", Proc. IEEE International symposium on antennas and propagation, Los Angeles, 1981, pp.683-686.

(c-11) Yamauchi, J., Nakano, H. and Iio, S., "Improvement of front-to-back ratio of backfire helical antenna", Proc. IEEE International symposium on antennas and propagation, Albuquerque, 1982, pp.370-373.

(c-12) Nakano, H. and Yamauchi, J., "Alternative derivation of integral equation for arbitrary antennas and scatterers", Proc. IEEE International symposium on antennas and propagation, Albuquerque, 1982, pp.592-595.

(c-13) Nakano, H., Yamane, T. and Yamauchi, J., "Directive phenomenon of parasitic helix and its application to antenna system", Proc. IEE Third International symposium on antennas and propagation, Norwich, 1983, pp.168-172.

(c-14) Nakano, H., Asaka, N. and Yamauchi, J., "Short helical antenna array fed from a waveguide", Proc. IEEE International symposium on antennas and propagation, Houston, 1983, pp.405-408.

(c-15) Yamauchi, J., Nakano, H. and Hashimoto, S., "Polarization diversity of 4-arm Archimedean spiral antenna", Proc. IEEE International symposium on antennas and propagation, Houston, 1983, pp.130-133.

(c-16) Nakano, H., Mikawa, T. and Yamauchi, J., "Numerical analysis of mono-filar conical helix", Proc. IEEE International symposium on antennas and propagation, Boston, 1984, pp.177-180.

(c-17) Nakano, H. "Simplified integral equation on electromagnetic coupling between arbitrarily bent wires and slots", Proc. IEEE International symposium on antennas and propagation, Boston, 1984, pp.413-416.

(c-18) Nakano, H., Yamauchi, J., Eda, M. and Iwasaki, T., "Numerical analysis of electromagnetic couplings between wires and slots using integral equations", Proc. IEE Fourth International symposium on antennas and propagation, Coventry, 1985, pp.438-442.

(c-19) Nakano, H., Yamauchi, J. and Yoshizawa, A., "The moment method for electromagnetic coupling between arbitrarily bent wire and slot structures", Proceedings of IECE International conference on antennas and propagation, Kyoto Japan, 1985, pp.851-854.

(c-20) Nakano, H., Mimaki, H. and Yamauchi, J., "Numerical analysis of a helical antenna with a finite ground plane", Proc. IEEE International symposium on antennas and propagation, Philadelphia, 1986, pp.129-132.

(c-21) Nakano, H., Tanabe, M., Yamauchi, J. and Shafai, L., "Spiral slot antenna", Proc.IEE Fifth International conference on antennas and propagation, York, 1987, pp.86-89.

(c-22) Nakano, H., Minegishi, Y. and Yamauchi, J., "Dual spiral antenna", Proc. IEEE International symposium on antennas and propagation, Virginia, 1987, pp.352-355.

(c-23) Nakano, H., Kerner, S. R. and Alexopoulos, N. G., "The moment method solution for printed antennas of arbitrary configuration", Proc. IEEE International symposium on antennas and propagation, Virginia, 1987, pp.1016-1019. (Note that a misprint of $k^2=\omega^2\sqrt{\mu_0\varepsilon_0}$ in Eq.(1) should be changed to $k=\omega\sqrt{\mu_0\varepsilon_0}$.)

Index

W

Z